I0480366

Contents

iii

Preface

Engineering students and all professionals working in any engineering sector require a complete understanding of Engineering Drawing. This book provides a comprehensive overview of Engineering Drawing in simple and practical oriented way supported with easy and well explained examples and exercises. It is intended to help undergraduate engineering students learn the fundamentals of Engineering Drawing.

The book covers all the basic contents covered in syllabus of Engineering Drawing or Technical Drawing courses in Undergraduate Engineering programs. It has 7 major chapters that contain all the relevant information required to understand the basics of Engineering Drawing. The book contains a large number of exercises at the end of each chapter to provide the students enough exercises for practice.

'**Learning by Practice**' is the basic principle of Engineering Drawing

Addisu Dagne Zegeye

CHAPTER 1

INTRODUCTION

1.1 What is Engineering Drawing?

Engineering Drawing is a graphic representation of a real thing. It is a universal graphic language of engineers. It is used by them to develop and record their ideas and transmit them to others for execution. It is the shortest of all short hands as it can express the complete information about an object say a machine part, with exactness and detail as no written language can describe even in many pages.

To draw something as a figure by means of lines expressing some ideas on a paper is the drawing. The purpose of the drawing is to define and specify the shape and size of a particular object by means of lines, other in formations about the object, which can not be expressed by lines, are given side by side on the drawing in a simplest and shortest way. A good type of drawing gives full information about the object in a shortest and simplest way. Hence drawing is the shortest of shorthands.

A drawing worked out by an engineer, having engineering ideas, for engineering purpose, is an Engineering drawing. It is the universal graphic language of engineers, a world language, a language of use and ever increasing value. It is spoken, read and written in its own way. Every language has its own rules of grammar. Engineering drawing also has been devised according to certain rules. It has its grammar in the theory of projection, its idioms in the conventionalized practices, its punctuations in the type of lines, its abbreviations, symbols, and its description in the construction. As a bad language is unpleasant to a master in the language, a wrong drawing will worry a trained eye of drawing. We have to learn to write a language so that we may be able to read it. If we know how to draw a drawing, we will be able to read and explain it. The knowledge of the drawing is the most important requirement of all the technical persons to work in an engineering occupation.

Machine drawing is one of the parts of engineering drawing pertaining to the drawing of machines. Mechanical engineers are mainly concerned with machine drawing.

1

Technical drawing is also a graphic language rightly applied to a drawing used to express technical ideas.

Sketching is the freehand expression of the graphic language. Sketching is the most important tool for the engineers engaged in technical work. Technical ideas can be expressed quickly and effectively by the sketches without the use of instruments.

1.2 History of technical drawing

Perhaps the earliest known technical drawing in existence is the plan view for a design of a fortress drawn by the Chaldean engineer Gudea and engraved upon a stone table, figure 1.1. It is remarkable how similar this plan is to those made by modern architects, although "drawn" thousands of years before paper was invented.

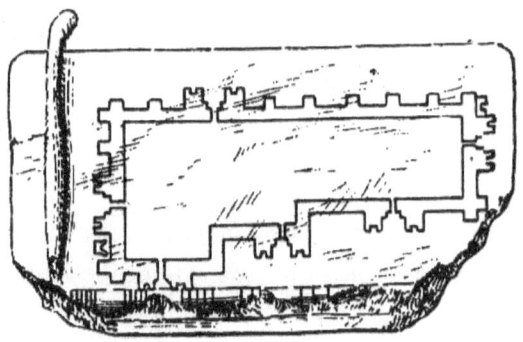

Fig. 1.1 Plan of a Fortress. This stone table is part of a statute now in the Louvre, Paris, and is classified in the earliest period of Chaldean art, about 4000 B.C.

In museums we can see actual specimens of early drawing instruments. Compasses were made of bronze and were about the same size as those in current use. As shown in fig. 1.2 the old compass resembled the dividers today.

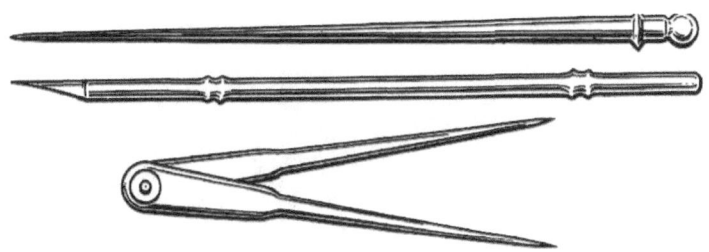

Fig. 1.2 Early drawing instruments

For upward of twenty thousand years, a drawing has been the principal means for the portrayal of ideas through the use of lines. Its beginnings, however, are still further back in time, for our early ancestors undoubtedly explained their ideas by making in the dust on the floors of their caves.

The earliest records of man are graphic, depicting people, deer, buffalo, and other animals of the time on the rock walls of caves. These drawings were to satisfy an elemental need for expression, long before the development of writing. However, drawing gradually freed itself from this early usage when writing was developed and it then came to be used primarily by artists and engineering designers as a means of setting forth ideas for the construction of finished works such as pyramids, war chariots, buildings, and simple mechanisms useful to man. Most of the very early drawings that still exist were made on parchment, which was very durable. Later, during the twelfth century, paper was developed in Europe and came into general use for drawings.

Only a few of the earliest drawings for fortresses, buildings, and simple mechanisms are in existence today. Those that have come down to us have been largely pictorial in nature, and they exist as carvings and paintings on walls of structures or have been woven in to tapestry. One of the earliest representations shows the use of the wheel about 3200 B.C. in Mesopotamia. The drawing presents a wheel barrow like structure being used by a man to transport his wife or child. The pictorial is primitive without any depth of perspective.

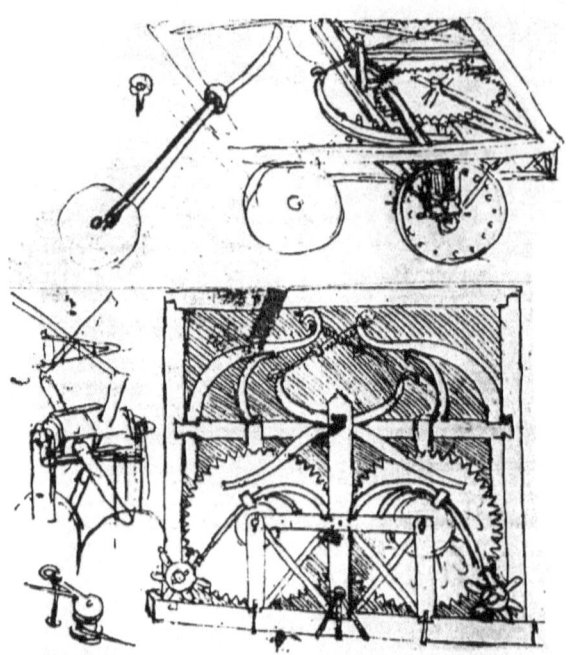

Fig. 1.3 Idea sketch prepared by Leonardo da Vinci (1452-1519). The da Vinci "automobile" was to have been powered by two giant springs and steered by the tiller, at the left in the picture, attached to the small wheel.

At the beginning of the Christian era, Roman architects had become skillful in preparing drawings that were to be constructed. They used straight edges and compasses to lay out the elevation and plan views and were able to prepare well-executed perspectives. However, the theory of projection of views upon imaginary planes was not developed as a means of representation until sometime during the Renaissance period. It is well known that Leonardo da Vinci used to record and transmit to others his ideas and designs for mechanical constructions, and many of these drawings are still in existence, Fig. 1.4.

Fig. 1.4 An Arsenal, by Leonardo da Vinci

Even though it is probable that Leonardo da Vinci was aware of the theory of multiview drawing his trainings as an artist prevailed and he recorded his ideas and designs for war machines and mechanical constructions by preparing perspective sketches and drawings such as the one shown in Fig.1.3. No multiview drawings prepared by da Vinci have been found.

1.3 Classification of Drawing

The drawing may be classified in to the following two distinct categories

1. Artistic Drawing
2. Engineering Drawing

Artistic drawing is the drawing or art of a person who draws sketches of a job by his imagination or keeping the job before him. The artist tries to produce the job in the shape of the picture. He is so perfect in keeping his art that he can prepare the picture of a job by imagination without measuring the size and picture looks quite proportionate. He requires only his pen or pencil and paper to prepare the picture. Dimensions and other details are not given in it; however, one can appreciate the shape and size of the job. It is not a simple drawing. One requires a great practice to prepare it. Everybody can understand and like this pictorial drawing.

Engineering drawing cannot be understood by every person; even the artist cannot understand it. It is the graphic language of engineers and those trained to read and write it can understand it. Dimensions and other details are also given in this drawing without which it is incomplete.

CHAPTER 2
THEORY OF PROJECTIONS

2.1 Introduction

In Engineering, Architectural and Construction works representing three-dimensional objects on two-dimensional paper has been a major obstacle to communicate ideas. The basic problem of representing these three dimensions' objects on a two-dimensional sheet of paper has been solved using the **concept of projection.**

In this concept, in order to represent an object by a two dimensional line drawing on a plane, imaginary '**Projectors**' emanating from various points on the object are extended until they pierce a picture plane. If we assume the projectors to be lines of sight from the object, our flat sheet of drawing paper becomes the plane of projection up-on which we represent the object. All projection theory is based on two variables, the line of sight and Plane of projection.

Line of Sight (LOS)

A line of sight is an imaginary ray of light between an observer's eye and an object.

Plane of projection

A plane of projection is an imaginary flat plane upon which the image created by the lines of sight is projected. The image is produced by connecting the points where the lines of sight pierce the projection plane. In effect, the 3-D object is transformed into a 2- D representation (also called a projection). The paper or computer screen on which a sketch or drawing is created is a plane of projection.

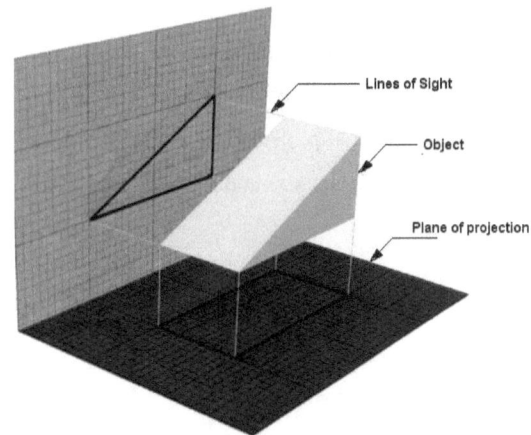

Figure 2.1 Plane of projection and lines of sight

7

2.2 Types of Projection

Projection is a drawing of a three-dimensional object on a two-dimensional surface. The two-dimensional surface used for projection is called *plane of projection* or the *picture plane*.

The classification of different types of projections depends on:

1. The angle the lines of sight (projection line) make with the projection plane
2. The angle the lines of sight make with each other.
3. The relative position of the object to be projected with respect to the projection plane.

Projection methods are broadly classified into two:

1. Central Projection (Perspective projection) and
2. Parallel projection

2.2.1　Central (perspective) projection

In central projection, the observer is assumed to be located at some finite location. Hence the visual rays projected from the different corners of the object converge to the single point of viewing. This is the actual viewing mechanism and therefore the projection possesses realistic appearance of the object. Perspective projection is not suitable for working drawings because a perspective view does not reveal exact size and shape.

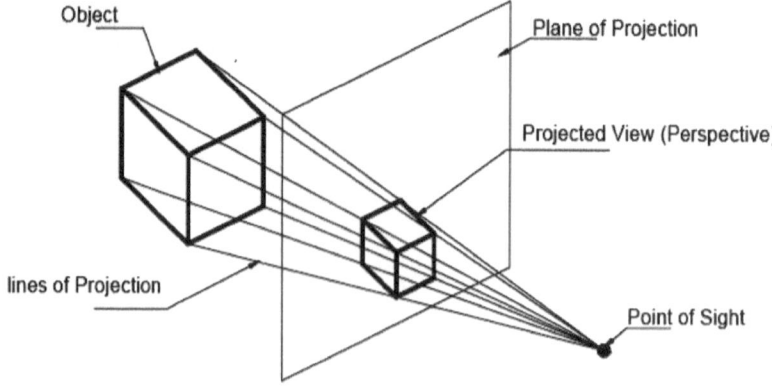

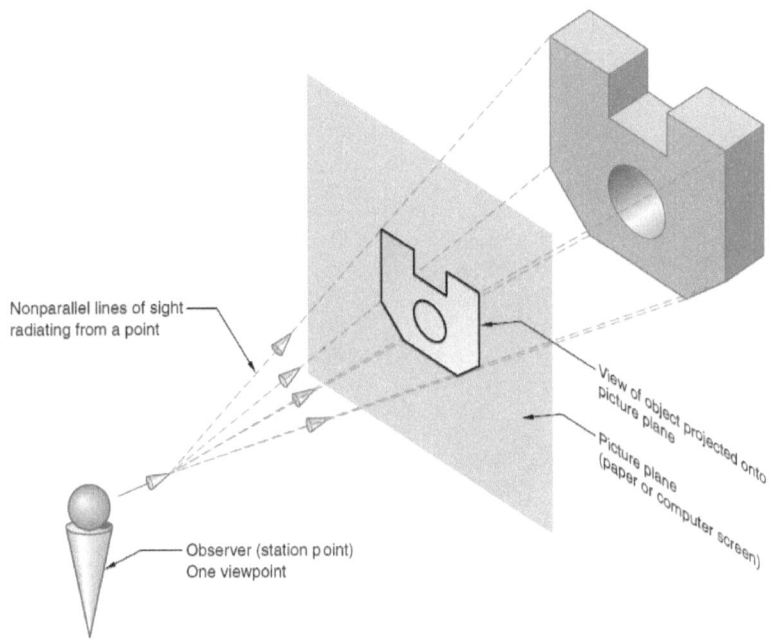

Nonparallel lines of sight radiating from a point

View of object projected onto picture plane

Picture plane (paper or computer screen)

Observer (station point)
One viewpoint

Fig 2.2 Central projection

2.2.2   Parallel Projection

If we can place the observer at very large distance from the object, the angle between the visual rays reflected from the different corners of the object becomes more and more gentle and thus we can assume that the visual rays become parallel to each other. Parallel projection is based on the assumption of observer being at large distance without losing the ability to see the object. This

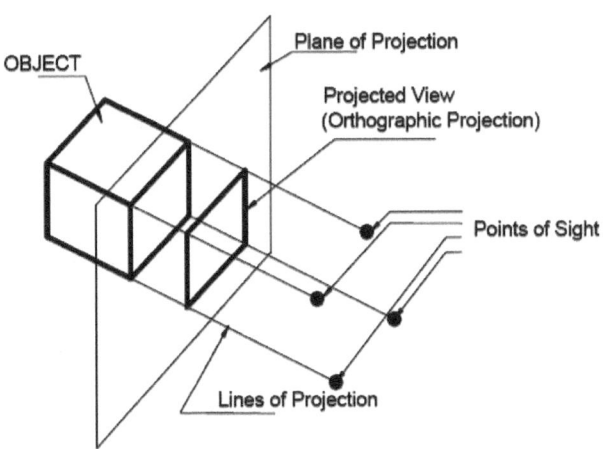

OBJECT

Plane of Projection

Projected View
(Orthographic Projection)

Points of Sight

Lines of Projection

9

assumption will bring about some distortion to the pictorial appearance of the projection as compared to actual appearance. Parallel projections are less realistic, but they are easier to draw.

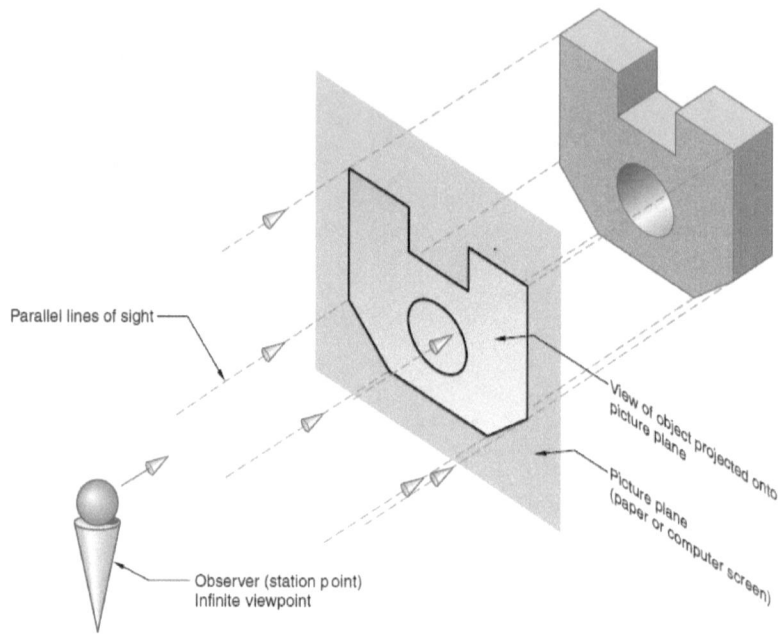

Parallel lines of sight

View of object projected onto picture plane

Picture plane (paper or computer screen)

Observer (station point)
Infinite viewpoint

Fig 2.3 Parallel projection

Parallel projection is further classified into *orthographic projection* and *oblique projection.* If the lines of sight are parallel to each other and perpendicular to the picture plane, the resulting projection is called an *orthographic projection.* If the lines of sight are parallel to each other but inclined to the picture plane, the resulting projection is called *oblique projection.*

Orthographic projection is a parallel projection technique in which the plane of projection is positioned between the observer and the object and is perpendicular to the parallel lines of sight. The orthographic projection technique can produce either pictorial drawing that show all the three dimensions of an object in one view or multi-views that show only two dimensions of an object in a single view. Orthographic projection is further classified into *multi-view projection* and *axonometric projection.* In multi-view projection more than one projection is used to give complete size and shape description of the object while in axonometric projection a single view is sufficient to describe the object completely.

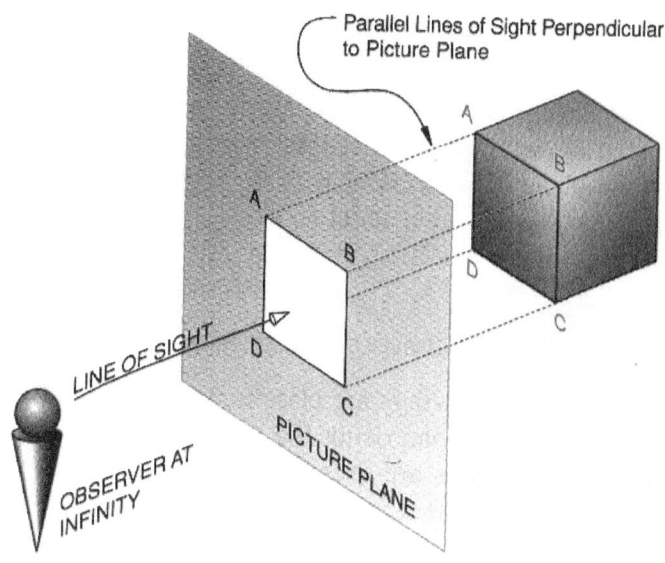

(A) Multiview Projection

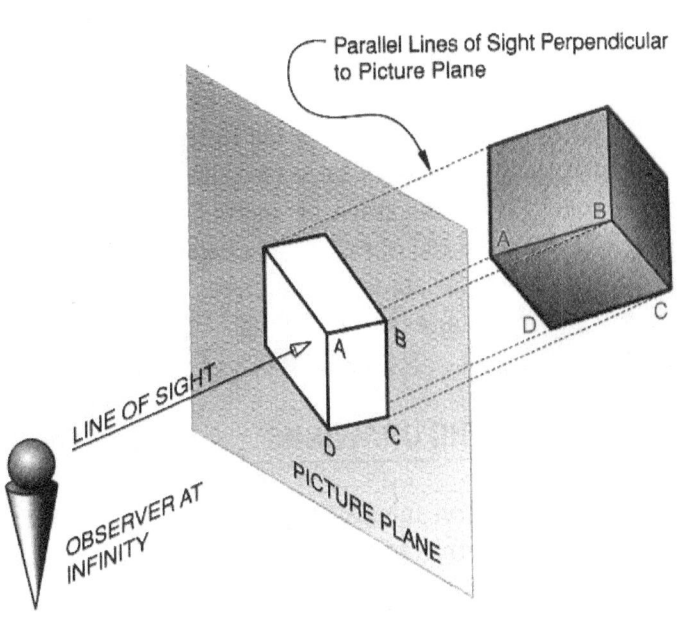

(B) Axonometric Projection

Fig. 2.4 Orthographic projection. A) Multiview Projection B) Axonometric Projection

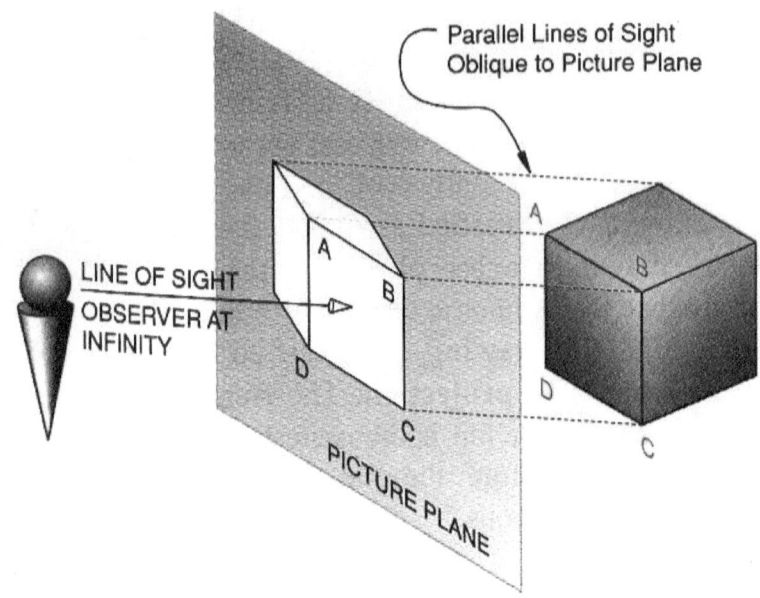

Fig. 2.5 Oblique parallel projection

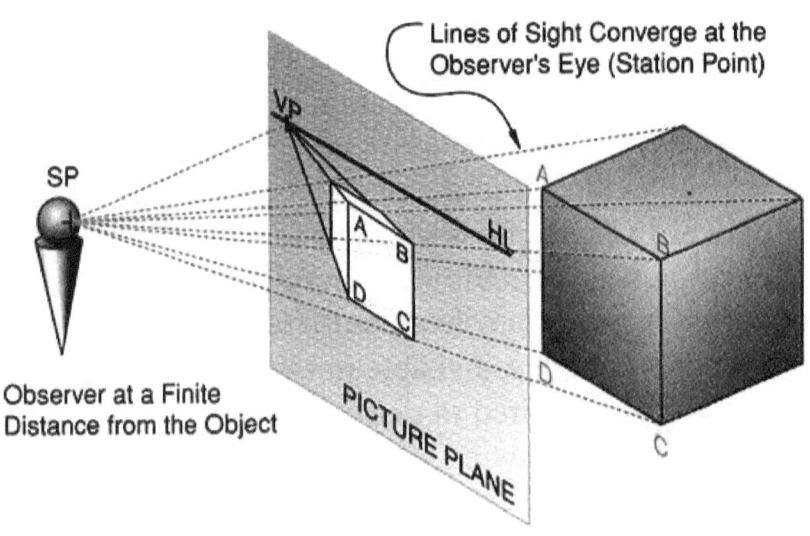

Fig. 2.6 Perspective (central) projection

2.3 Axonometric Projection

Axonometric projection is a parallel projection technique used to create a pictorial drawing of an object by rotating the object on an axis relative to a projection plane. In axonometric and multiview projections the lines of sight are perpendicular to the plane of projection (thus they are called Orthographic projection). The differences between a multiview drawing and an axonometric drawing are that, in a multiview, only two dimensions of an object are visible in each view and more than one view is required to define the object, whereas, in an axonometric drawing, the object is rotated about an axis to display all the three dimensions, and only one view is required.

2.3.1 Types of Axonometric Drawings

Axonometric drawings are classified by the angles between the lines comprising the axonometric axes. The **axonometric axes** are axes that meet to form the corner of the object that is nearest to the observer.

Isometric

Isometric drawings are the quickest and easiest of all the pictorials to draw and the most commonly used. In an isometric drawing the three normal surfaces of a rectangular solid will have equal angles between them (120°).

Dimetric

In dimetric drawings, two of the normal surfaces will be equally spaced, but the third surface will have an angle of a different number of degrees.

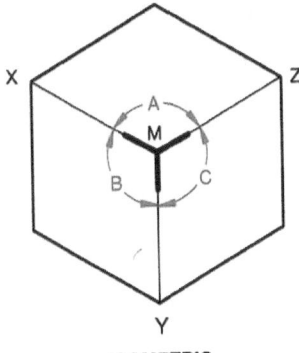

ISOMETRIC
Angles A, B, and C are equal
Edges MZ, MY, and MX are equal
in length

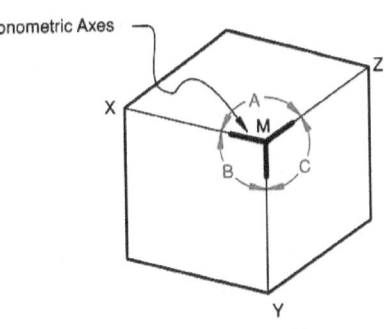

DIMETRIC
Angles A and C are equal
Edges MY and MX are equal in
length

Trimetric

A trimetric drawing will have the three normal surfaces of the rectangular solid positioned so none of the three angles have the same number of degrees. The drawing of dimetric and trimetric drawings takes more time because uncommon angles are often used.

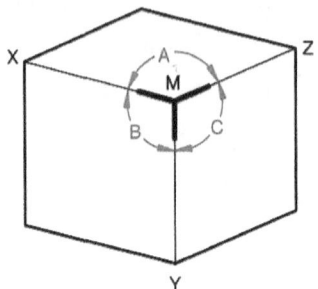

Trimetric: No equal edges and No equal Angles

2.4 Oblique Drawings

Oblique drawings are very similar to isometric drawings, but these are some noticeable differences. The object has one face of the rectangular solid parallel to the picture plane. Two other faces do appear but since the angle of the depth axis may vary, the angles of the other two surfaces may vary from drawing to drawing. The axes of an oblique drawing are horizontal, vertical and depth axis. You may select any angle for the depth axis but a 45^0 angle is usually preferred.

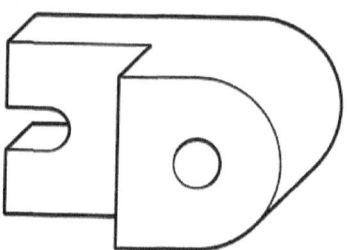

Fig. 2.7 Oblique Drawing

The main advantage of drawing an oblique drawing is that you may use the circle template on the plane that is parallel to the picture plane. This is possible because the circles will appear as true circles and not as ellipses on this surface only. Remember, any plane parallel to the picture plane is true size and shape.

2.5 Summary of projection theory

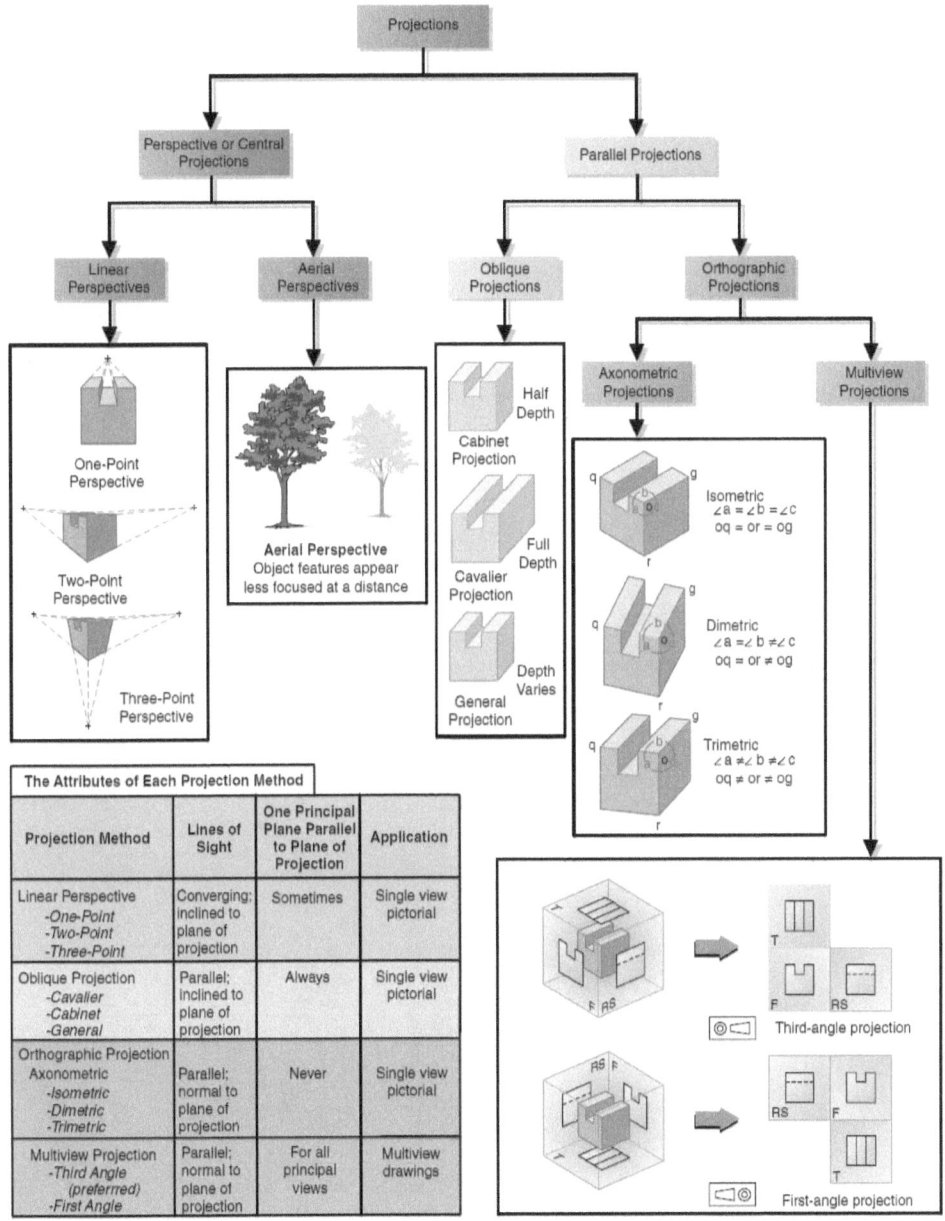

Fig. 2.8 Summary of projection techniques developed along two lines: parallel and perspective

15

CHAPTER 3
MULTIVIEW DRAWINGS

3.1 Introduction

Multiview projection is an orthographic projection for which the object is behind the plane of projection, and the object is oriented such that only two of its dimensions are shown. As parallel lines of sight pierce the projection plane, the features of the part are outlined.

Multiview drawings employ multiview projection techniques. In multiview drawings, generally three views of an object are drawn, and the features and dimensions in each view accurately represent those of the object. Each view is a 2-D flat image. The views are defined according to the positions of the planes of the projection with respect to the object.

3.2 Planes of Projection

Three planes, that are mutually perpendicular to each other, are the basis of multi-view projection. The plane in vertical position is known as the *vertical projection plane* or *frontal projection plane*. The one in the horizontal position is called the *horizontal projection plane*. The third plane which is perpendicular to the first two is called the *profile projection plane*. The horizontal and vertical planes are called the principal planes of projection.

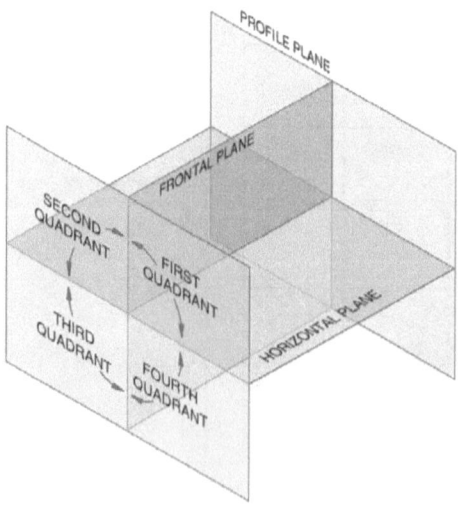

Fig 3.1 The three Principal Planes of projection

16

In multi-view projections, the lines of sight (projectors) are perpendicular to the planes of projection. The projection of an object on the vertical/frontal projection plane is called vertical/frontal projection. The vertical projection of an object is commonly known as front view. Similarly, horizontal projection of an object is called top view and the profile projection of an object is known as side view.

3.2.1 Frontal plane of projection

The frontal plane of projection is the plane onto which the front view of a multiview drawing is projected. The front view of an object shows the width and height dimensions.

3.2.2 Horizontal plane of projection

The top view of an object is projected onto the horizontal plane of projection, which is a plane suspended above and parallel to the top of the object. The top view of an object shows the width and depth dimensions.

3.2.3 Profile plane of projection

The right side view of an object is projected onto the right profile plane of projection, which is a plane that is parallel to the right side of the object. The side view of the object shows the depth and height dimensions. In multiview drawings, the right side view is the standard side view used.

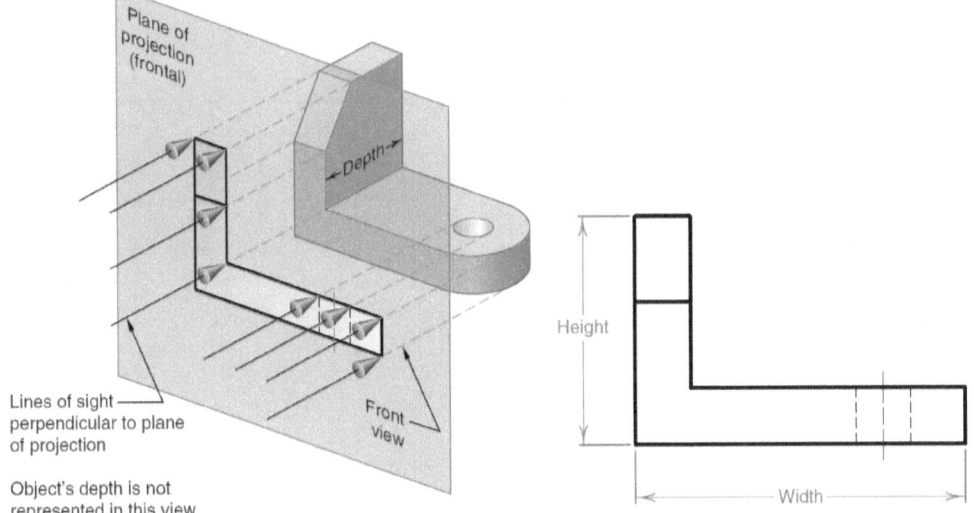

Fig 3.2 Front view

17

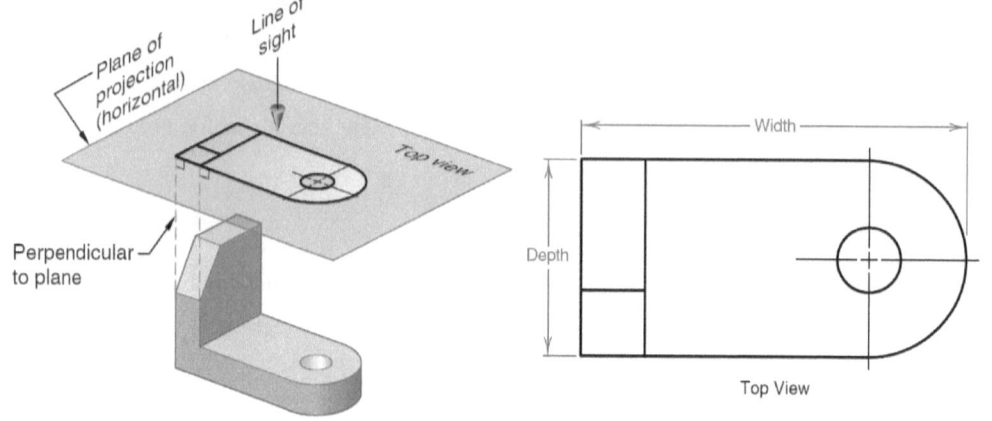

Fig 3.3 Top view

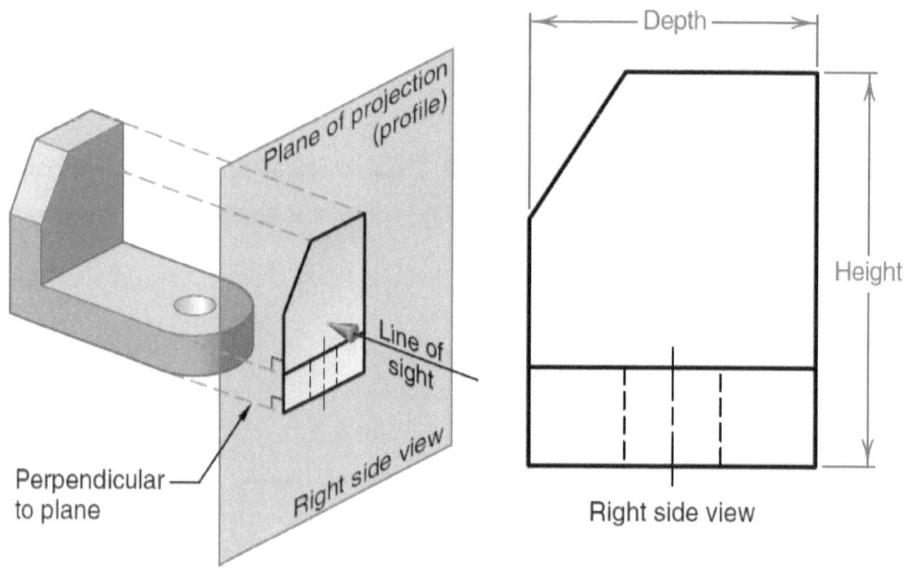

Fig 3.4 The Right side view

3.3 First and Third Angle Projections

As illustrated in fig 3.1, the three principal planes of projection form four quadrants. The three projections of an object can be made by placing the object in one of the four quadrants. However, only the first and the third quadrants are conventionally used. If the object is placed in the first

18

quadrant, the resulting projection is called **first angle projection**. Similarly, if the object is placed in the third quadrant, the resulting projection is called **third angle projection**. Third angle projection is used in the United States, Canada, and many other countries throughout the world. First angle projection is mainly in European and Asian countries. We will use third angle projection throughout this course.

3.3.1 Third-Angle Projection

The third-angle projection method is an orthographic representation in which the object to be represented and seen by the observer appears behind the coordinate viewing planes on which the object is orthographically projected. i.e. the <u>projection plane is placed between the observer and the object</u>. Therefore, in the projection process it is necessary to assume that the plane of projection is transparent. On each projection plane, the object is represented as if it is seen orthogonally from in front of each plane.

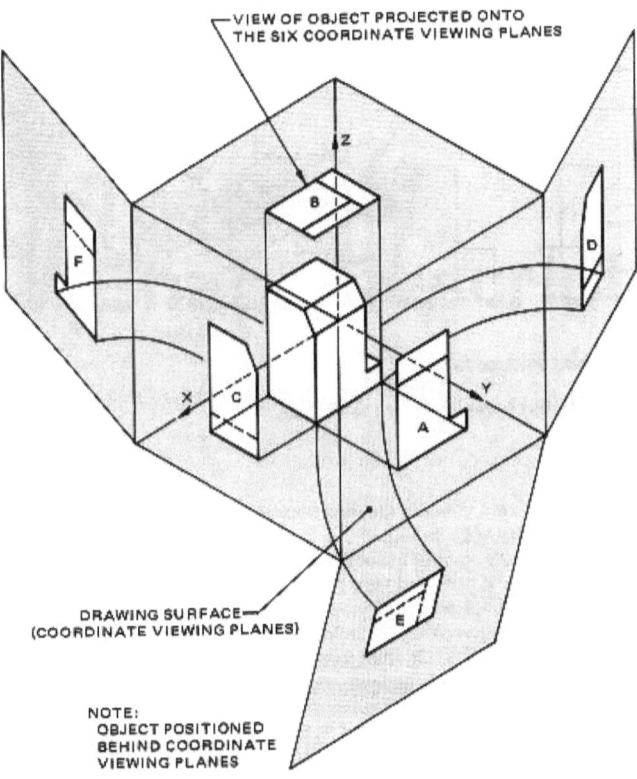

Fig 3.5 Third angle projection

19

Figure 3.5 illustrates the method of getting the third angle projections of an object. After developing the different views of the object in the different projection planes, all the projection planes are opened up to the frontal projection plane. Note that, in third angle projection system, the top view is placed directly above the front view and the right side view is placed directly to the right of the front view. Figure 3.6 below shows positioning of views on a drawing surface in third angle projection method.

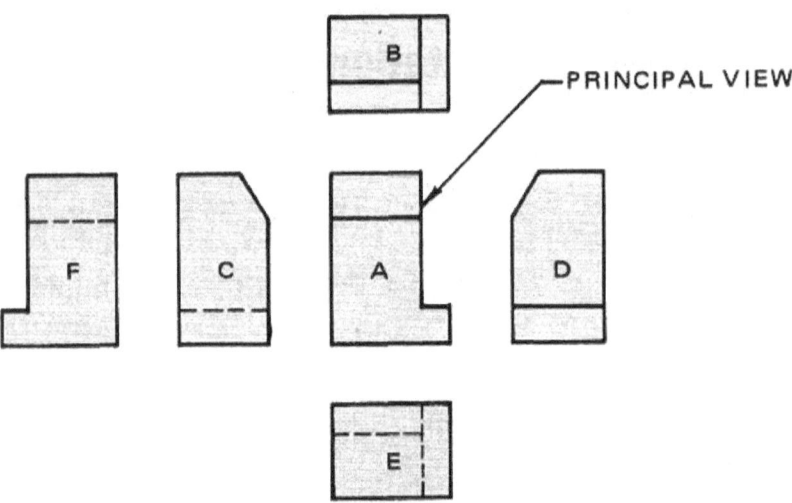

Fig 3.6 Positioning of views on a drawing surface in third angle projection method

In figure 3.6 shown above, View A is Front view, View B is Top View, View C is Left side view, View D is Right side view, View E is Bottom view, View F is Rear side view

The symbol of Third angle projection is shown below

IDENTIFYING SYMBOL

3.3.2   First Angle projection

In the first angle projection system, the object is assumed to be placed in the first quadrant. If the observer is placed in front of the vertical plane, the object will appear to be between the observer and the projection plane.

The first angle projection method is an orthographic representation in which the object to be represented appears between the observer and the coordinate viewing planes on which the object is orthogonally projected.

In first angle projection the views are arranged in such a way that the top view is always directly below the front view and the left side view is directly to the right of the front view.

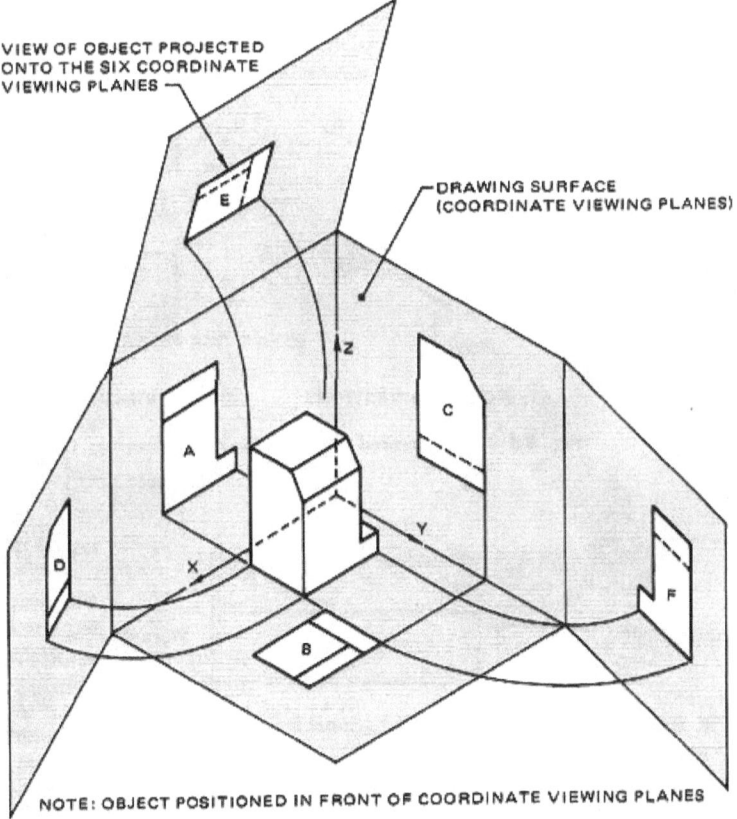

Fig 3.7 First angle projection

21

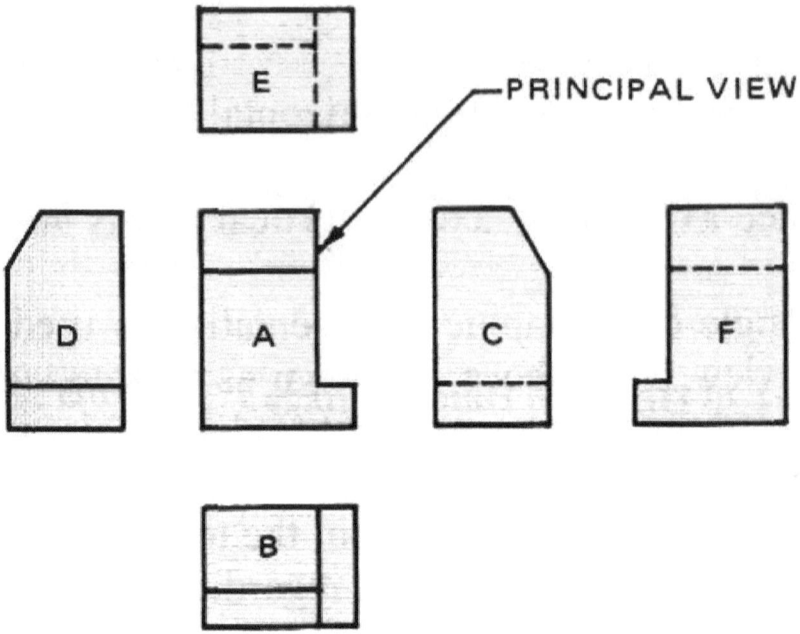

Fig 3.8 Positioning of views on a drawing surface in first angle projection method

Similarly, View A is Front view, View B is Top View, View C is Left side view, View D is Right side view, View E is Bottom view, View F is Rear side view

The symbol of first angle projection is

IDENTIFYING SYMBOL

3.4 The six principal views

The plane of projection can be oriented to produce an infinite number of views of an object. However, some views are more important than others. These principal views are the six mutually perpendicular views that are produced by six mutually perpendicular planes of projection.

If you imagine suspending an object in a glass box with major surfaces of the object positioned so that they are parallel to the sides of the box, the six sides of the box become projection planes showing the six views.

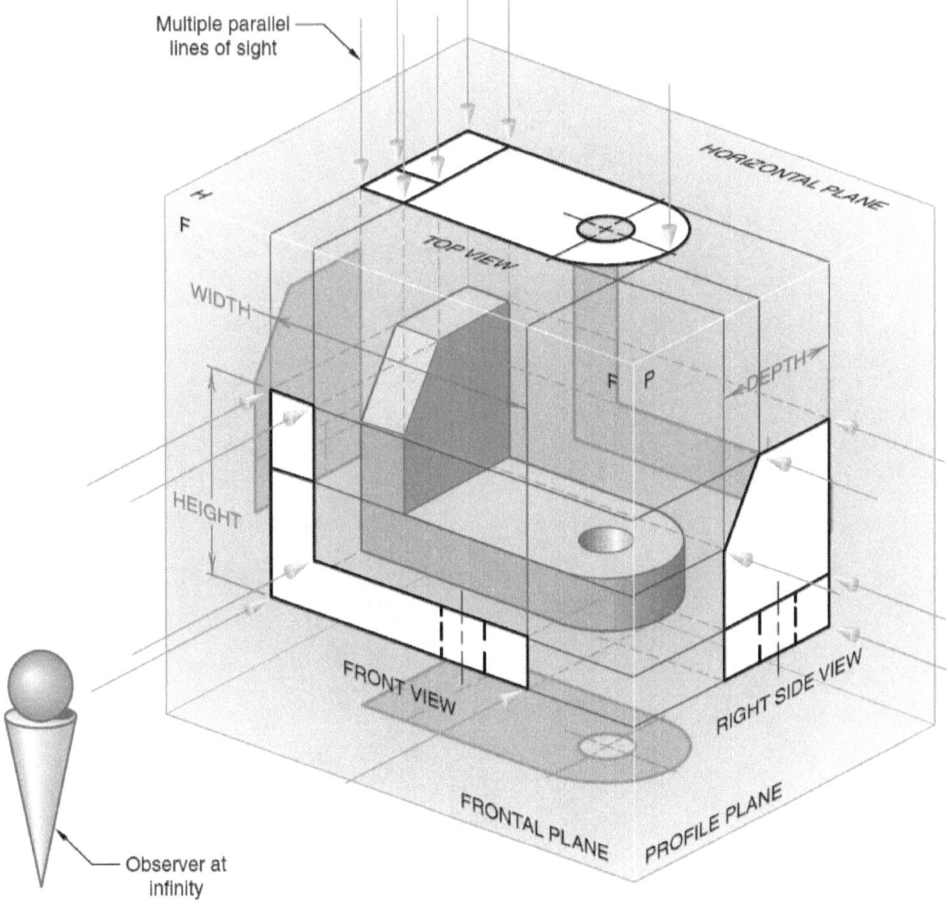

Fig 3.9 Object suspended in a Glass Box, producing the six principal Views

Each view is perpendicular to and aligned with the adjacent views. The six principal views are **front, top, left side, right side, bottom, and rear**. To draw these views on a 2-D media, that is, a piece of paper or a computer monitor, imagine putting hinges on all sides of the front glass plane and on one edge of the left profile plane. Then cut along all the other corners, and flatten out the box to create a six-view drawing as shown in figs 3.10 and 3.11.

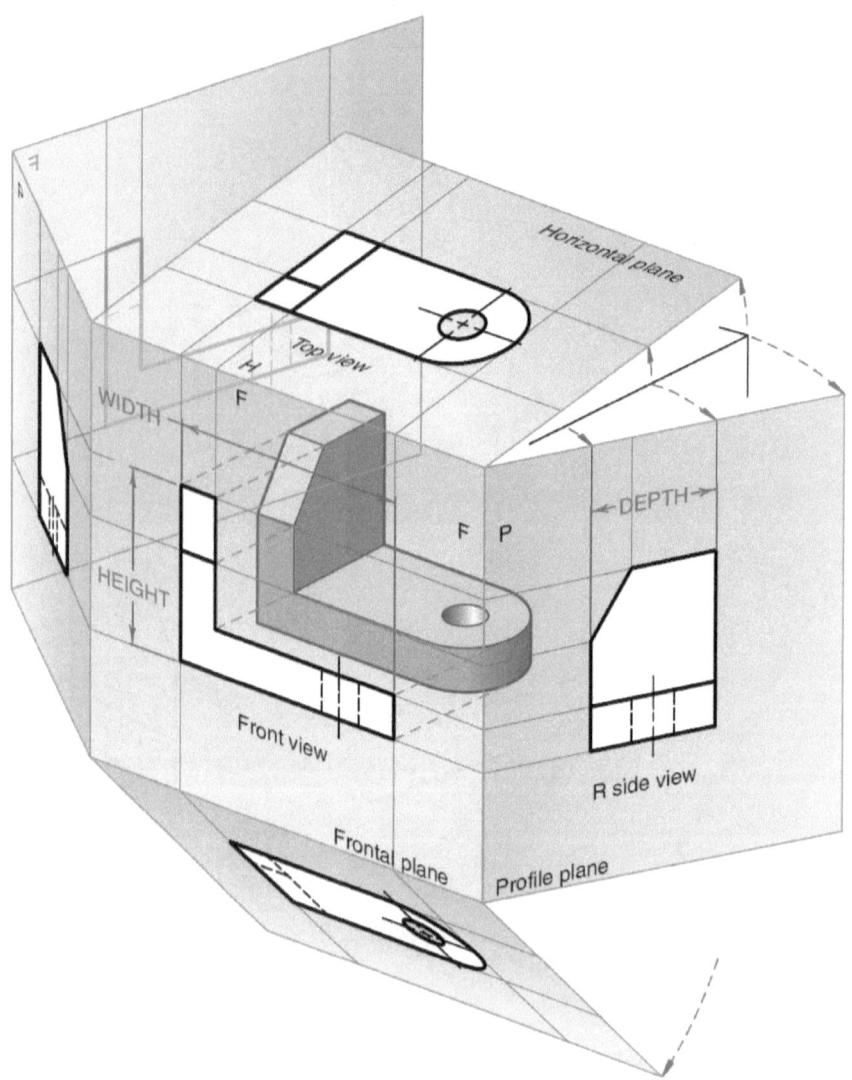

Fig 3.10 Unfolding the Glass box

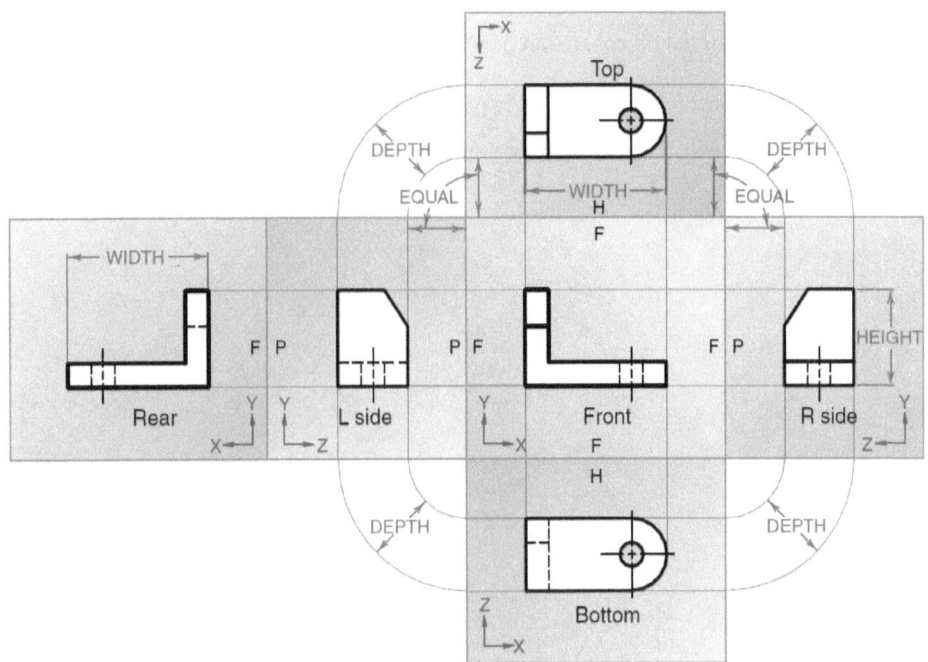

Fig 3.11 Six-view drawings produced when unfolding the glass box

The **front view** is the one that shows the most features or characteristics. All other views are based on the orientation chosen for the front view. Also, all other views, except the rear view, are formed by rotating the lines of sight 90 degrees in an appropriate direction from the front view.

The **right side view** shows what becomes the right side of the object once the position of the front view is established. The **left side view** shows what becomes the left side of the object once position of the front view is established. The left side view is a mirror image of the right side view, except that hidden lines may be different.

3.5 Orientation of the object

Before you start making views of an object, you should place it in space in accordance with the following rules:

- Place the object in its most natural (stable) position. The object must be positioned within the imaginary glass box such that the surfaces of major features are either perpendicular or parallel to the glass planes (Figure 3.12). This will create views with

25

a minimum number of hidden lines. Figure 3.13 shows an example of poor positioning: the surfaces of the object are not parallel to the glass.

- Place the object with its <u>main faces parallel to the planes of projection</u>.
- Place the object so that the <u>most complex and descriptive face is on the front side.</u>

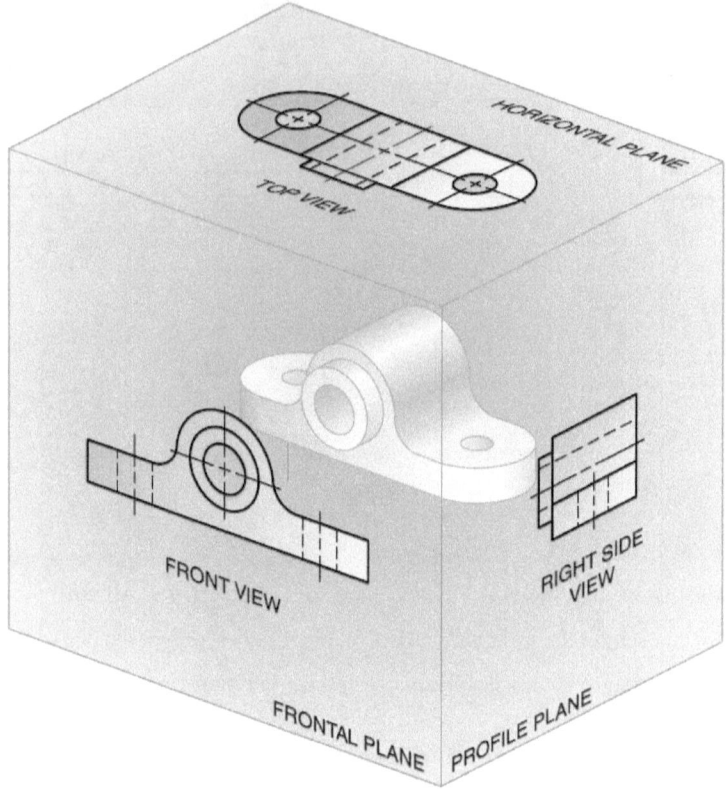

Fig 3.12 Good Orientation: suspend the object in a glass box such that major surfaces are parallel or perpendicular to the sides of the box (projection planes).

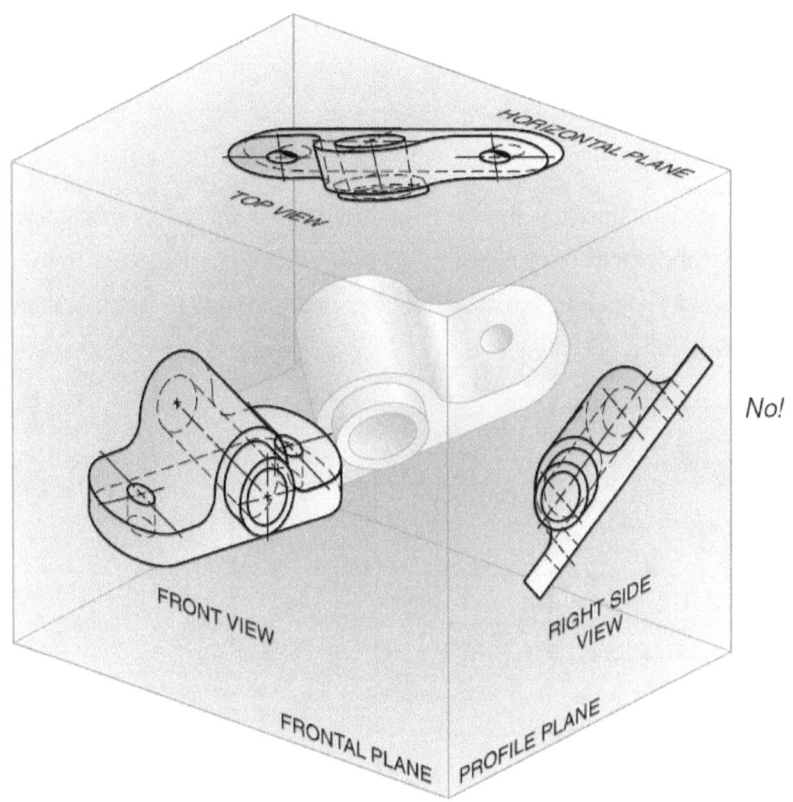

Fig 3.13 Poor Orientation: suspending the object so that surfaces are not parallel to the sides of the glass box produces views with many hidden lines

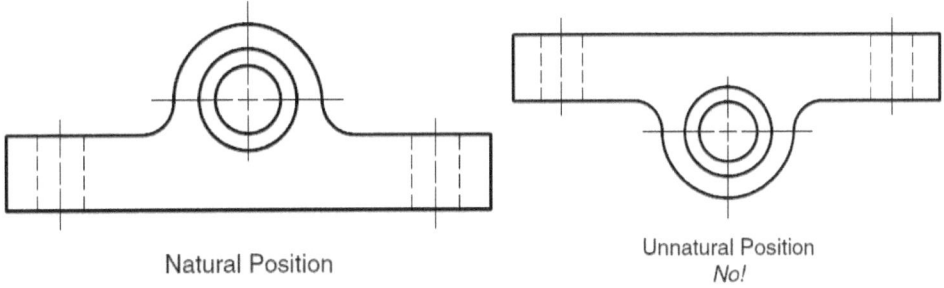

Natural Position

Unnatural Position
No!

Fig. 3.14 Natural position, always try to draw objects in their natural position

3.6 Choice of views

Once the orientation of the object is specified, the front direction, and thus the front view, is determined automatically. Then, in order to select the other necessary views, the following guidelines shall be used.

➤ Select those views (in addition to the front view) that provide the clearest information about the shape of the object. Select those views which are the most descriptive and have the fewest hidden lines. In the example the right side view has fewer hidden lines than the left side view.

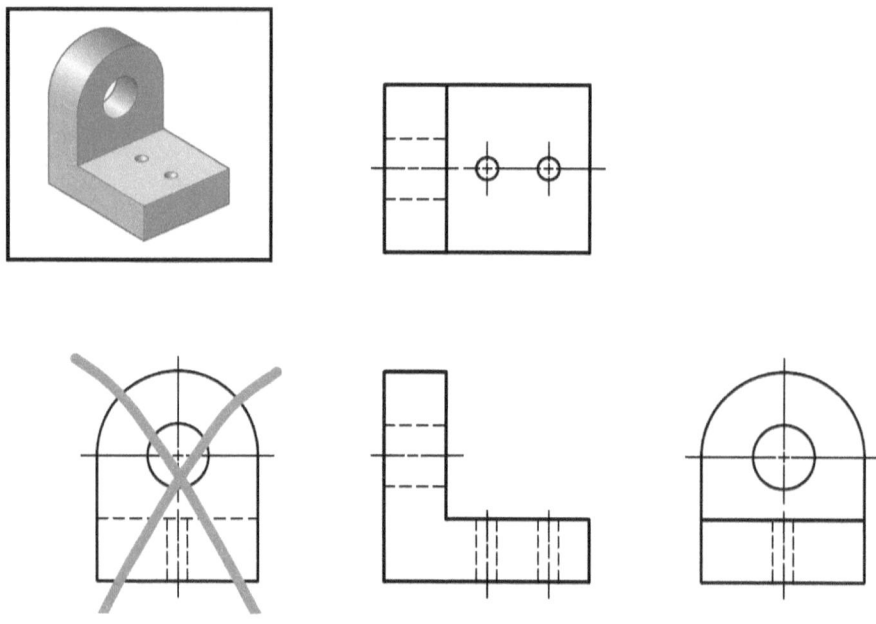

Fig 3.15 Most Descriptive Views

➤ Do not use more views than are necessary to describe the object. Select the minimum number of views needed to completely describe an object. Eliminate views that are mirror images of other views.

28

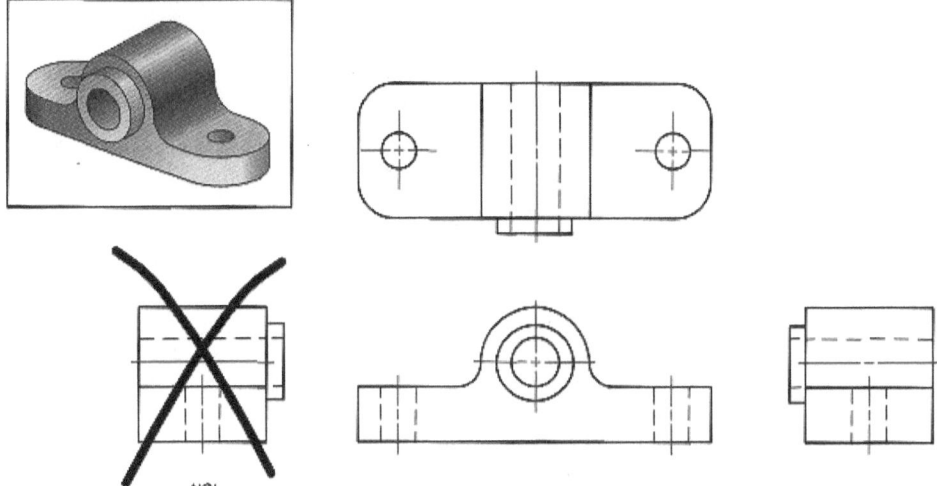

Fig 3.16 Minimum number of views

> If the left side and Right side views are identical in terms of information and line work, select the view to be drawn to the right of the front view, in accordance with tradition.

> If the top view and the bottom view are identical in terms of information and line work, select the top view.

> If the top and side views are identical in terms of information and line work, select the one that best utilizes the available drawing space.

> If two views are identical in terms of information, but one contains more hidden line work than the other, select the view with fewer hidden lines.

3.7 Three-View Drawings

The three-view multiview drawing is the standard used in engineering and technology, because many times the other principal views are mirror images and do not add the knowledge about the object. The standard views used in a three-view drawing are the **top, front, and right side views**, arranged as shown in figure below. The width dimensions are aligned between the front and the top views, using vertical projection lines. The height dimensions are aligned between the front and profile views, using horizontal projection lines. Because of the relative positioning of the three views, the depth dimension cannot be aligned using projection lines. Instead, the depth dimension is measured in either the top or right side view and transferred to the other view, using either a scale, miter line, compass, or dividers.

29

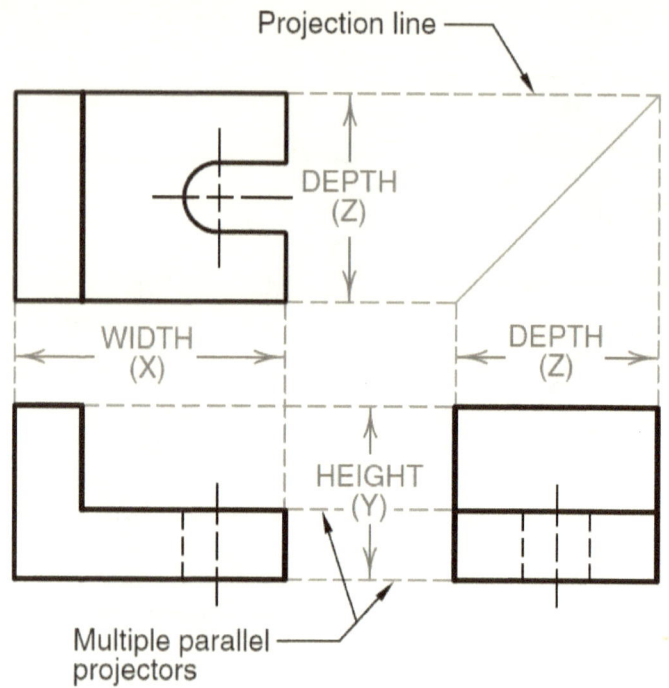

Fig. 3.17 Alignment of three view drawings

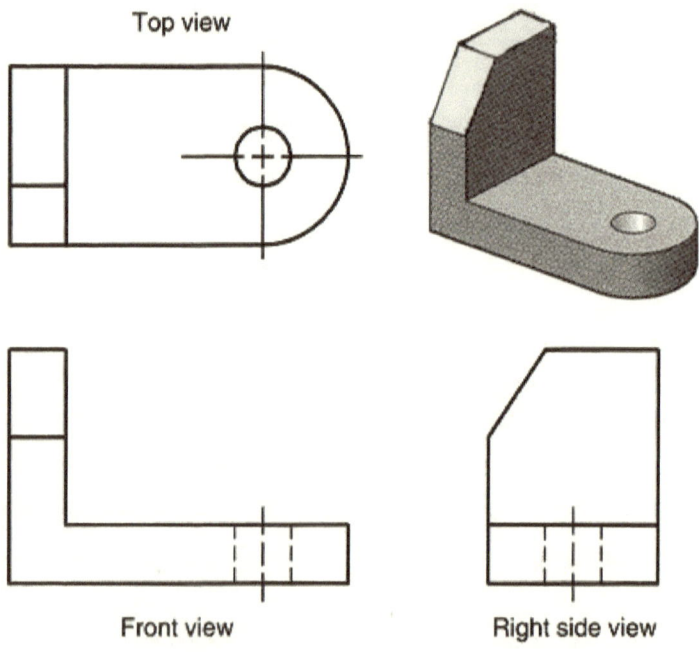

Fig. 3.18 Example of Multi-view drawing of an object

3.8 Laying out of a three-view drawing

To lay out a three-view drawing on a given drawing space, the three views should be spaced as illustrated below.

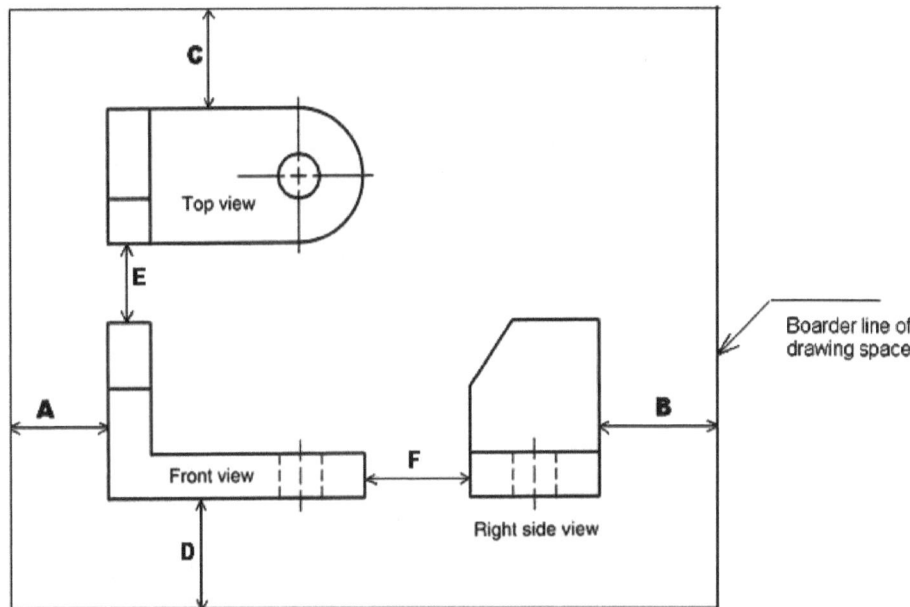

Fig. 3.19 Layout of a three-view drawing

Note that the length A and B should be equal, length C and D should be equal and length E and F should be equal. However, E or F will be set depending on the available space and appearance.

3.9 One-and Two-View Drawings

Some objects can be adequately described with only one view. A sphere can be drawn with one view because all views will be a circle. A cylinder or cube can be described with one view if a note is added to describe the missing feature or dimension. Other applications include a thin gasket or a printed circuit board.

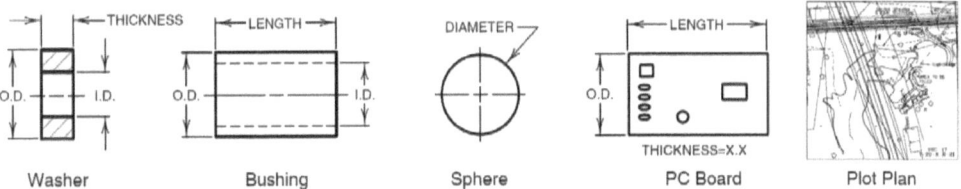

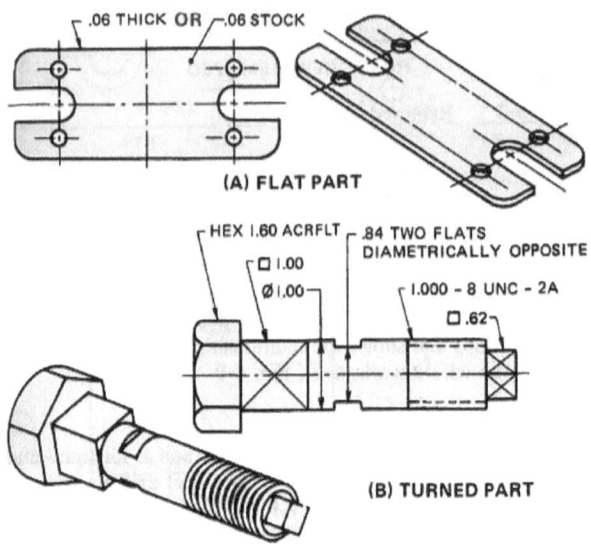

Fig. 3.20 One-view drawings, applications for one-view drawings include some simple cylindrical shapes, spheres, thin (flat) parts, and map drawings.

Other objects can be adequately described with **two views**. Cylindrical, conical, and pyramidal shapes are examples of such objects. For example, a cone can be described with a front and a top view. A profile view would be the same as the front view.

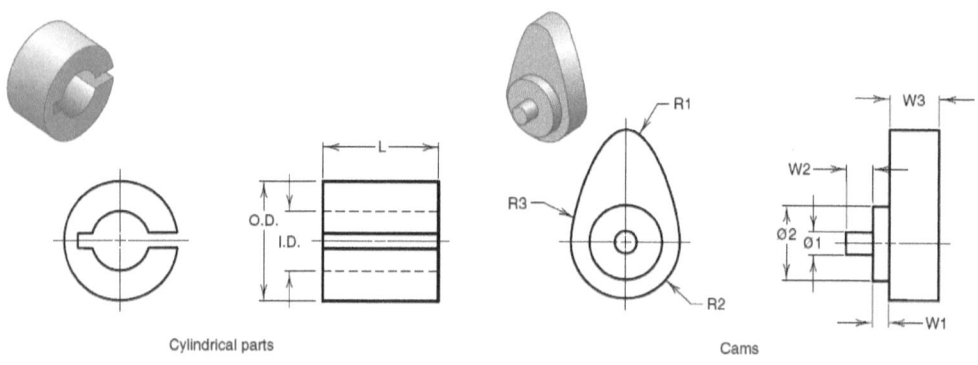

32

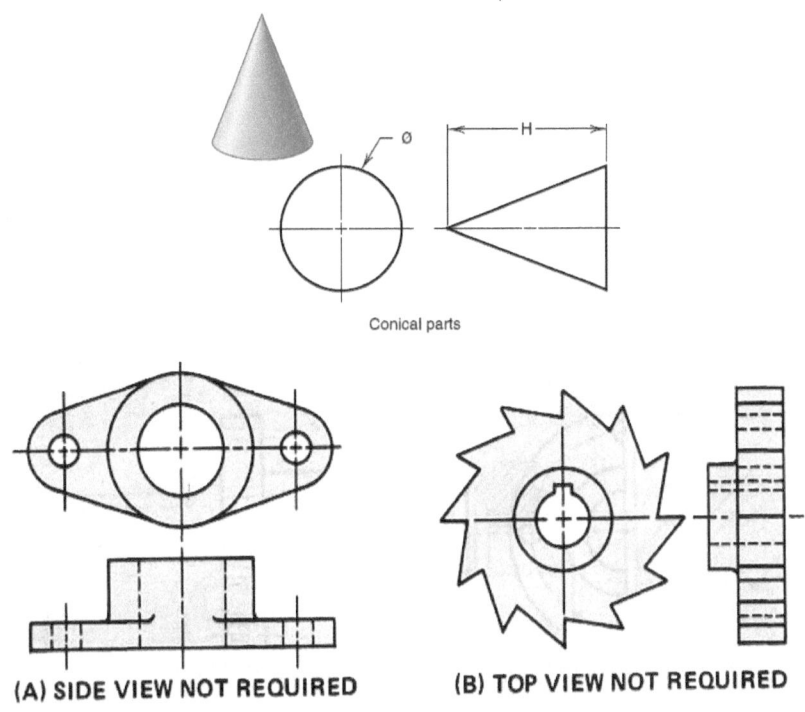

Conical parts

(A) SIDE VIEW NOT REQUIRED   (B) TOP VIEW NOT REQUIRED

Fig. 3.21 Two-view drawings, applications for one-view drawings include cylindrical and conical shapes

3.10   Use of a Miter line

The use of a miter line provides a fast and accurate method of constructing the third view once two views are established.

Using a Miter line to construct the Right Side View

1. Given the top and front views, project lines to the right of the top view.
2. Establish how far from the front view the side view is to be drawn (Distance D)
3. Construct the miter line at 45^0 to the horizontal.
4. Where the horizontal projection lines of the top view intersect the miter line, drop vertical projection lines.
5. Project horizontal lines to the right of the front view, and complete the side view

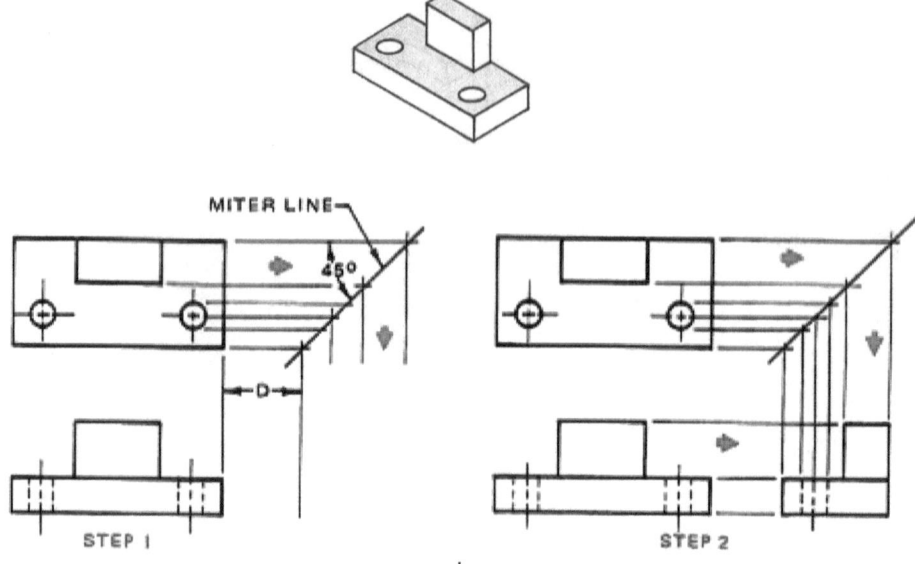

Fig. 3.22 Use of miter line to construct right side view from the given top and front views

Using a miter line to construct the top view

1. Given the front and side views, project vertical lines up from the side view.
2. Establish how far away from the front view the top view is to be drawn (Distance D)
3. Construct the miter line at 450 to the horizontal.
4. Where the vertical projection lines of the side view intersect the miter line, project horizontal lines to the left.
5. Project vertical lines up from the front view, and complete the top view.

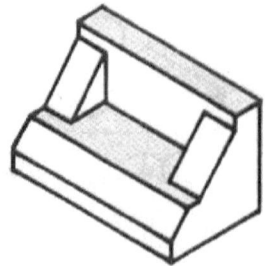

34

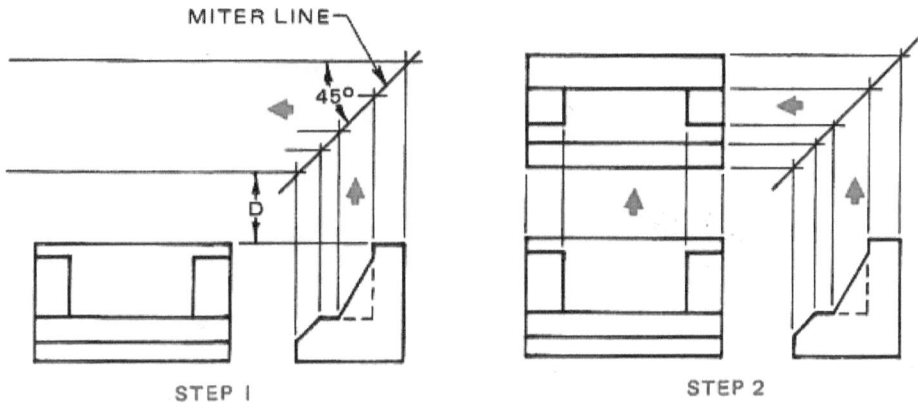

Fig. 3.22 Use of miter line to construct top view from a given front and side views

3.11 Fundamental views of edges and planes

In multiview drawings, there are fundamental views for edges and planes. These fundamental views show the edges or planes in true size, not foreshortened, so that true measurements of distances, angles, and areas can be made.

Edges (lines)

An edge is the intersection of two planes and is represented as a line on multiview drawings. A normal line, or true-length line, is an edge that is parallel and thus perpendicular to the line of sight. Edge 1-2 in fig. 3.23 in the top and right side view is a normal edge.

An edge appears <u>as a point</u> in a plane of projection to which it is <u>perpendicular</u>. Edge 1-2 is a point in the front view. The edge appears as a point because it is parallel to the line of sight used to create the front view.

An **inclined line** is parallel to a plane of projection but inclined to the adjacent planes, and it appears foreshortened in the adjacent planes. Line 3-4 is inclined and foreshortened in the top and right side view, but is true length in the front view because it is parallel to the front plane of projection

35

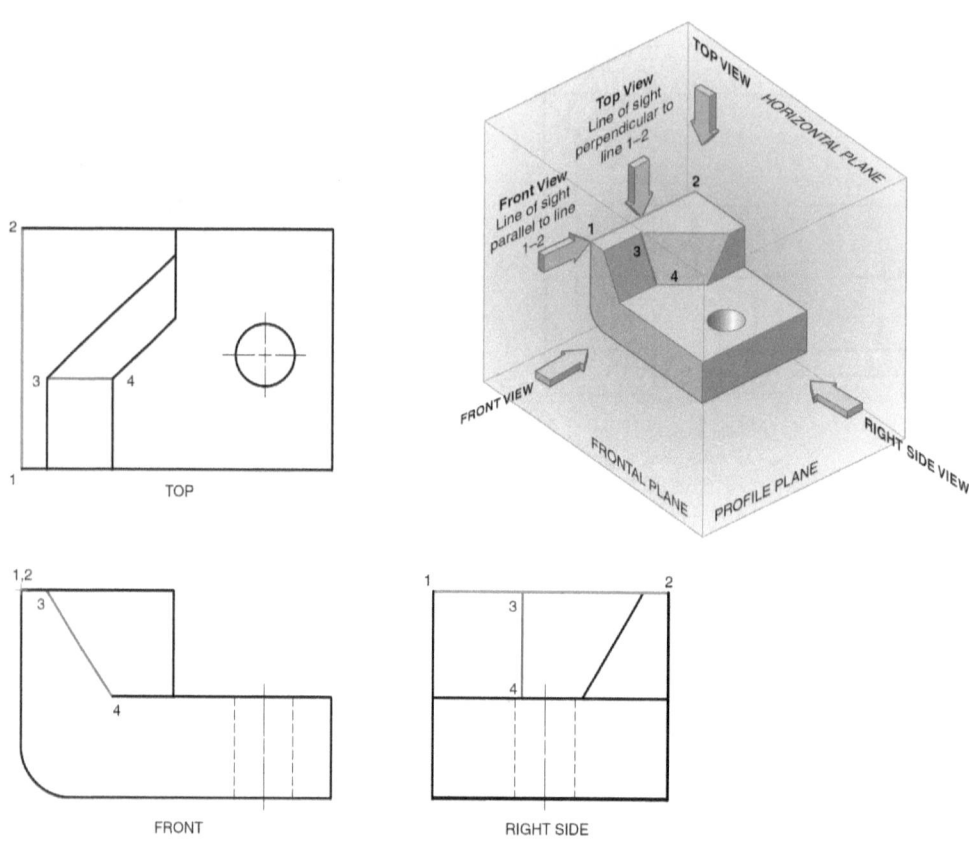

Fig. 3.23 Fundamental views of edges

An **oblique line** is <u>not parallel to any principal plane of projection</u>; therefore, it never appears as a point or in true length in any of the six principal views. Instead, an oblique edge will be foreshortened in every view and will appear as an inclined line. Line 1-2 in figure below is an oblique edge.

36

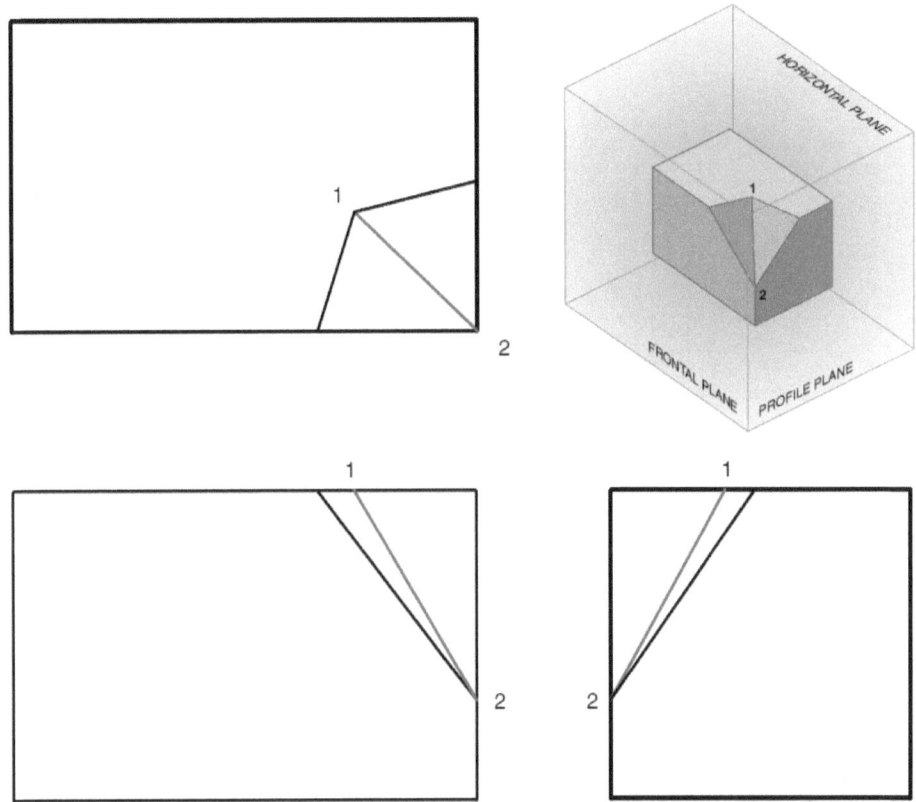

Fig. 3.24 Oblique line, oblique line 1–2 is not parallel to any of the principal planes of projection of the glass box

Principal Planes

A **principal plane** is parallel to one of the principal planes of projection and therefore perpendicular to the line of sight. A principal plane or surface will be true size and shape in the view where it is parallel to the projection plane and will appear as a horizontal or vertical line in the adjacent views. In figure below surface A is parallel to the frontal projection plane and is therefore a principal plane. Because surface A appears true size and shape in the front view, it is sometimes referred to as a **normal plane.** In this figure, surface A appears as a horizontal edge in the top view and as a vertical edge in the right side view.

A **frontal plane** is parallel to the front plane of projection and is true size and shape in the front view. A frontal plane appears as a horizontal edge in the top view and a vertical edge in the profile views. Surface A is a frontal plane.

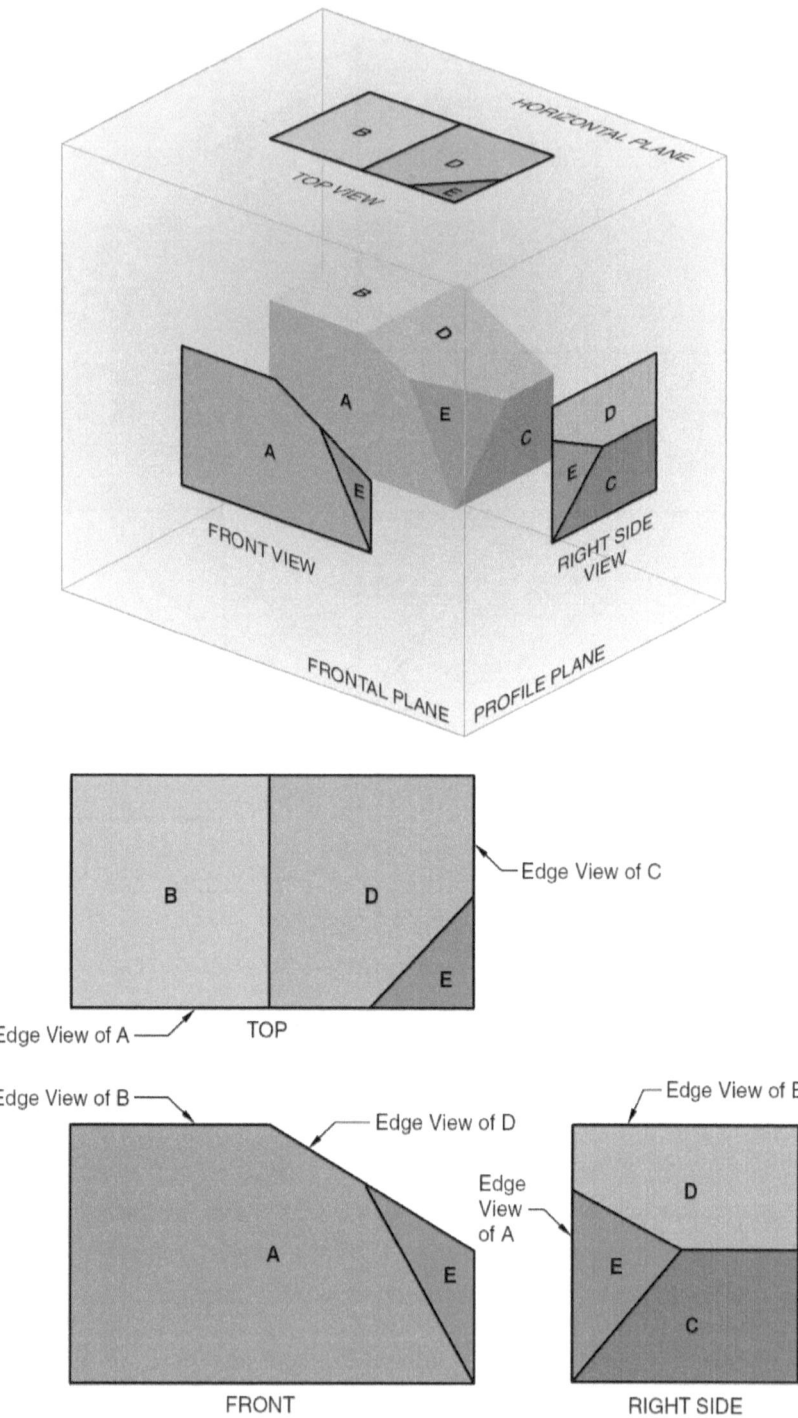

Fig. 3.25 Fundamental views of surfaces

A **horizontal plane** is parallel to the horizontal planes of projection and is true size and shape in the top (and bottom) view. A horizontal plane appears as a horizontal edge in the front and side views. In figure above Surface B is a horizontal plane

A **profile plane** is parallel to the profile (right or left side) planes of projection and is true size and shape in the profile views. A profile plane appears as a vertical edge in the front and top views. In figure above Surface C is a profile plane.

Inclined Surfaces

An **inclined Surface** is perpendicular to one of the projection and inclined to adjacent planes and can not be viewed in true size and shape in any of the principal views. An inclined surface appears as an edge in the view where it is perpendicular to the projection plane and as a foreshortened surface in the adjacent views. In Figure 3.25 above surface D is an inclined surface.

If the surfaces of an object lie in either a horizontal or a vertical position, the surfaces appear in their true shapes in one of the three views and appear as a line in the other two views.

When a surface is inclined or sloped in only one direction, that surface is not seen in its true shape in the top, front, or side view. It is, however, seen in two views as a distorted surface. On the third view it appears as a line. The true lengths of surfaces A and B in the figure below is seen in the front view only. In the top and side views, only the width of surfaces A and B appears in its true size. The length of these surfaces is foreshortened.

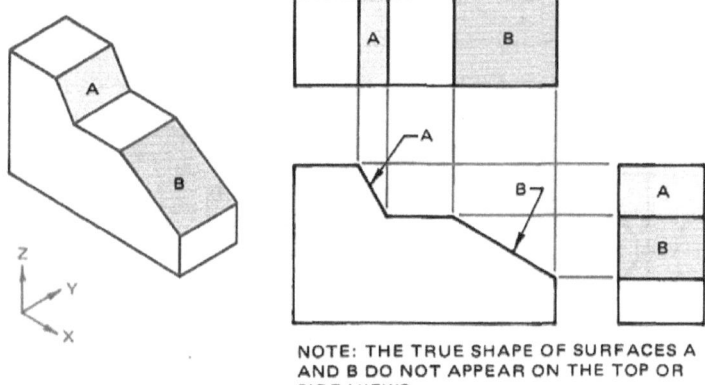

NOTE: THE TRUE SHAPE OF SURFACES A AND B DO NOT APPEAR ON THE TOP OR SIDE VIEWS.

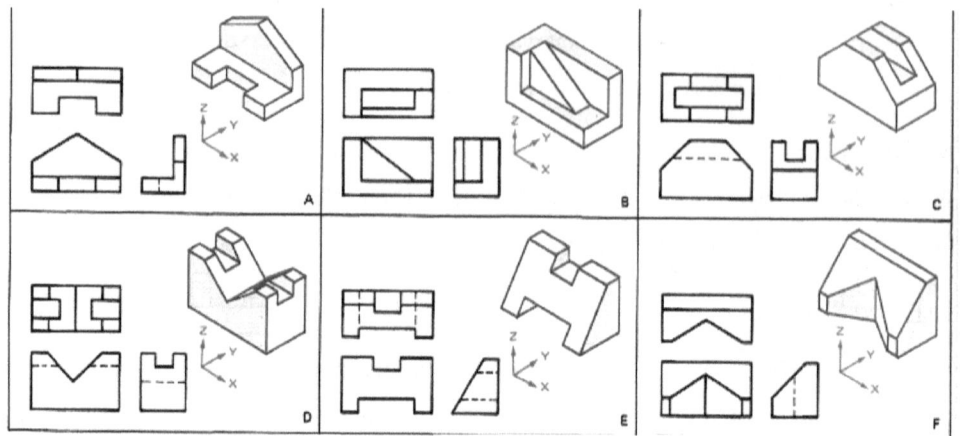

Fig. 3.26 Illustrations of objects having inclined surfaces

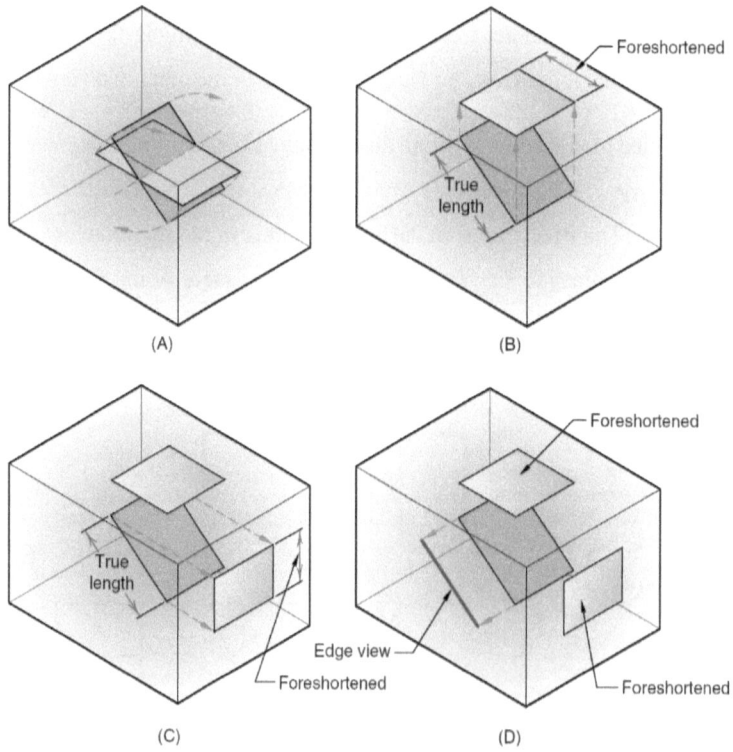

Fig. 3.27 Inclined face projection, an inclined face is oriented so that it is not parallel to any of the principal image planes. The inclined face is foreshortened in two views and is an edge in one view.

Oblique Surfaces

An oblique surface is not parallel to all the principal planes of projection. In figure 3.25 above, Surface E is an oblique surface, and in figure 3.28 below surface A is an oblique surface. An oblique surface does not appear in its true size and shape, or as an edge, in any of the principal views; instead, an oblique surface always appears as a foreshortened plane in the principal views.

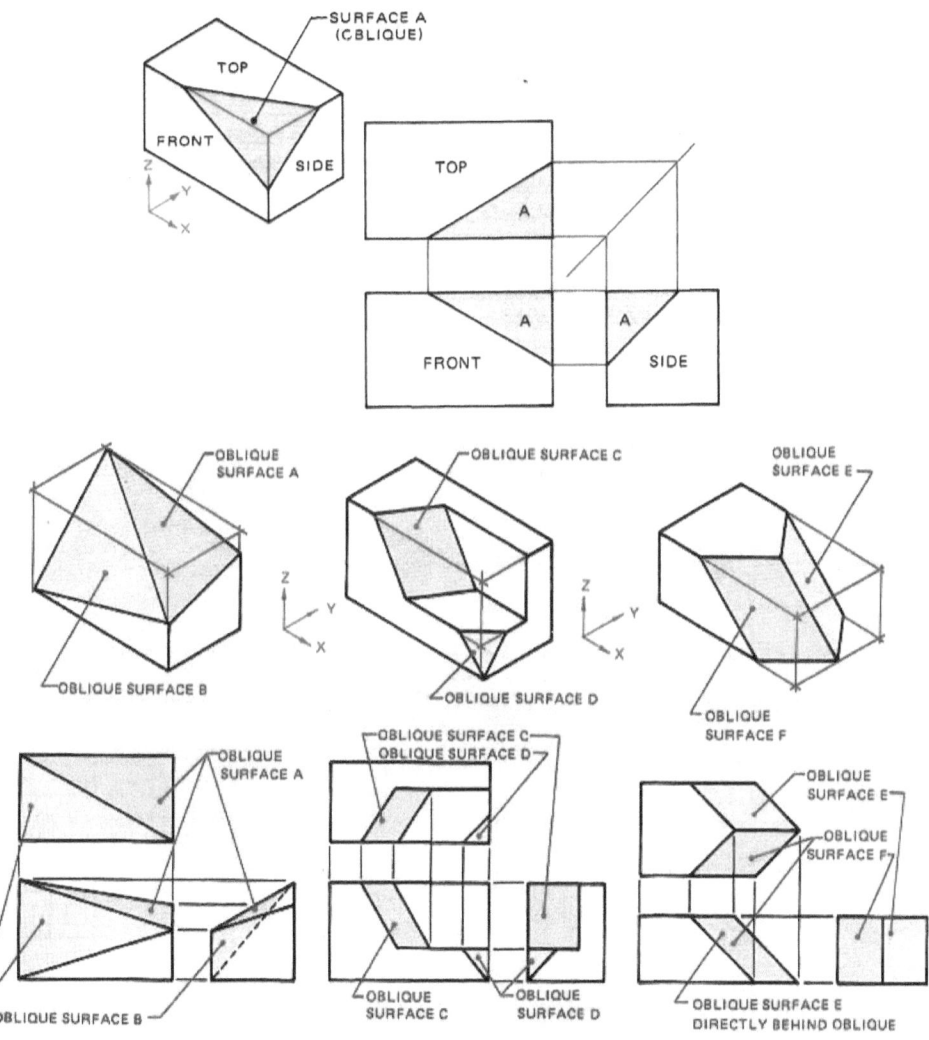

Fig 3.28 Illustration of objects having oblique surfaces

41

3.12 Line conventions

As in all engineering and technical drawings, multiview drawings and sketches require adherence to the proper use of the alphabet of lines.

TYPE OF LINE	APPLICATION	DESCRIPTION
HIDDEN LINE — — — — — THIN — — — — —		THE HIDDEN OBJECT LINE IS USED TO SHOW SURFACES, EDGES, OR CORNERS OF AN OBJECT THAT ARE HIDDEN FROM VIEW.
CENTER LINE THIN ALTERNATE LINE AND SHORT DASHES	CENTER LINE	CENTER LINES ARE USED TO SHOW THE CENTER OF HOLES AND SYMMETRICAL FEATURES.
SYMMETRY LINE CENTER LINE THICK SHORT LINES	SYMMETRY LINE	SYMMETRY LINES ARE USED WHEN PARTIAL VIEWS OF SYMMETRICAL PARTS ARE DRAWN. IT IS A CENTER LINE WITH TWO THICK SHORT PARALLEL LINES DRAWN AT RIGHT ANGLES TO IT AT BOTH ENDS.
EXTENSION AND DIMENSION LINES THIN DIMENSION LINE EXTENSION LINE		EXTENSION AND DIMENSION LINES ARE USED WHEN DIMENSIONING AN OBJECT.
LEADERS ARROW DOT THIN		LEADERS ARE USED TO INDICATE THE PART OF THE DRAWING TO WHICH A NOTE REFERS. ARROWHEADS TOUCH THE OBJECT LINES WHILE THE DOT RESTS ON A SURFACE.
BREAK LINES THIN LONG BREAK THICK SHORT BREAK		BREAK LINES ARE USED WHEN IT IS DESIRABLE TO SHORTEN THE VIEW OF A LONG PART.

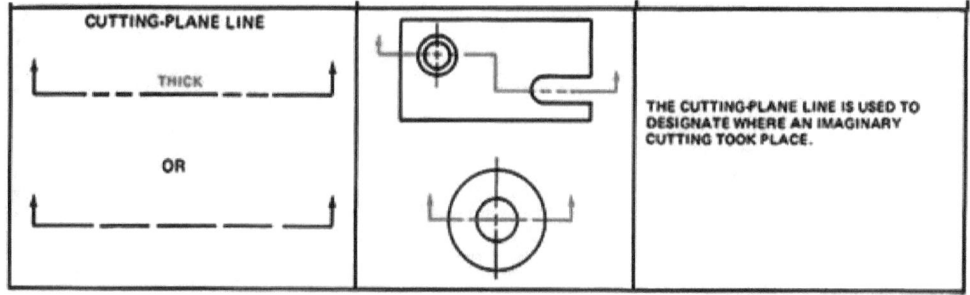

CUTTING-PLANE LINE

THICK ————————

OR

THE CUTTING-PLANE LINE IS USED TO DESIGNATE WHERE AN IMAGINARY CUTTING TOOK PLACE.

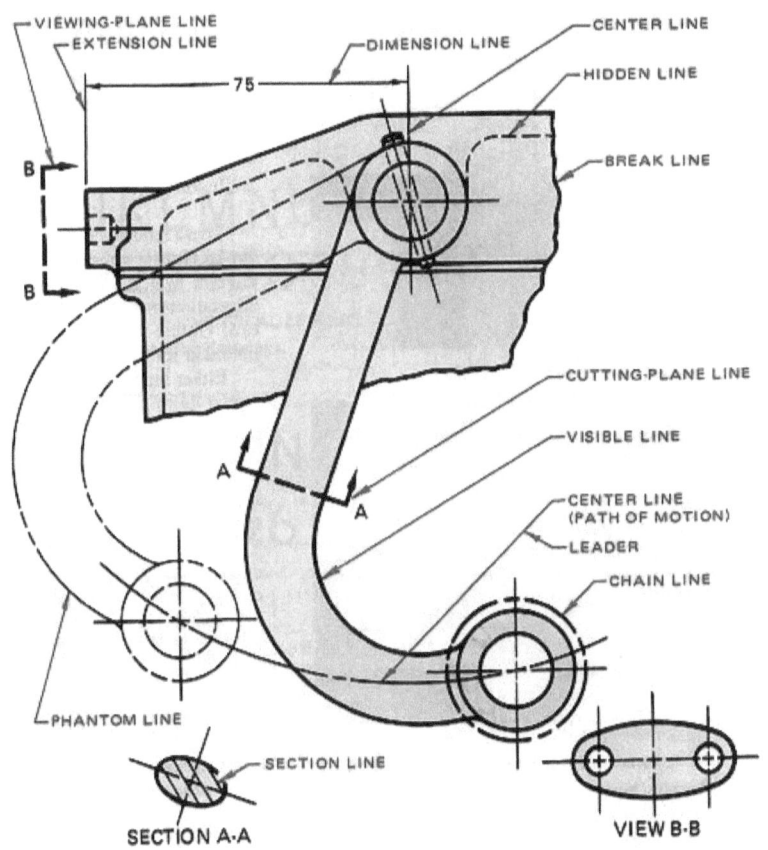

VIEWING-PLANE LINE
EXTENSION LINE
DIMENSION LINE
CENTER LINE
HIDDEN LINE
75
BREAK LINE
B
B
CUTTING-PLANE LINE
A
VISIBLE LINE
A
CENTER LINE
(PATH OF MOTION)
LEADER
CHAIN LINE
PHANTOM LINE
SECTION LINE
SECTION A-A
VIEW B-B

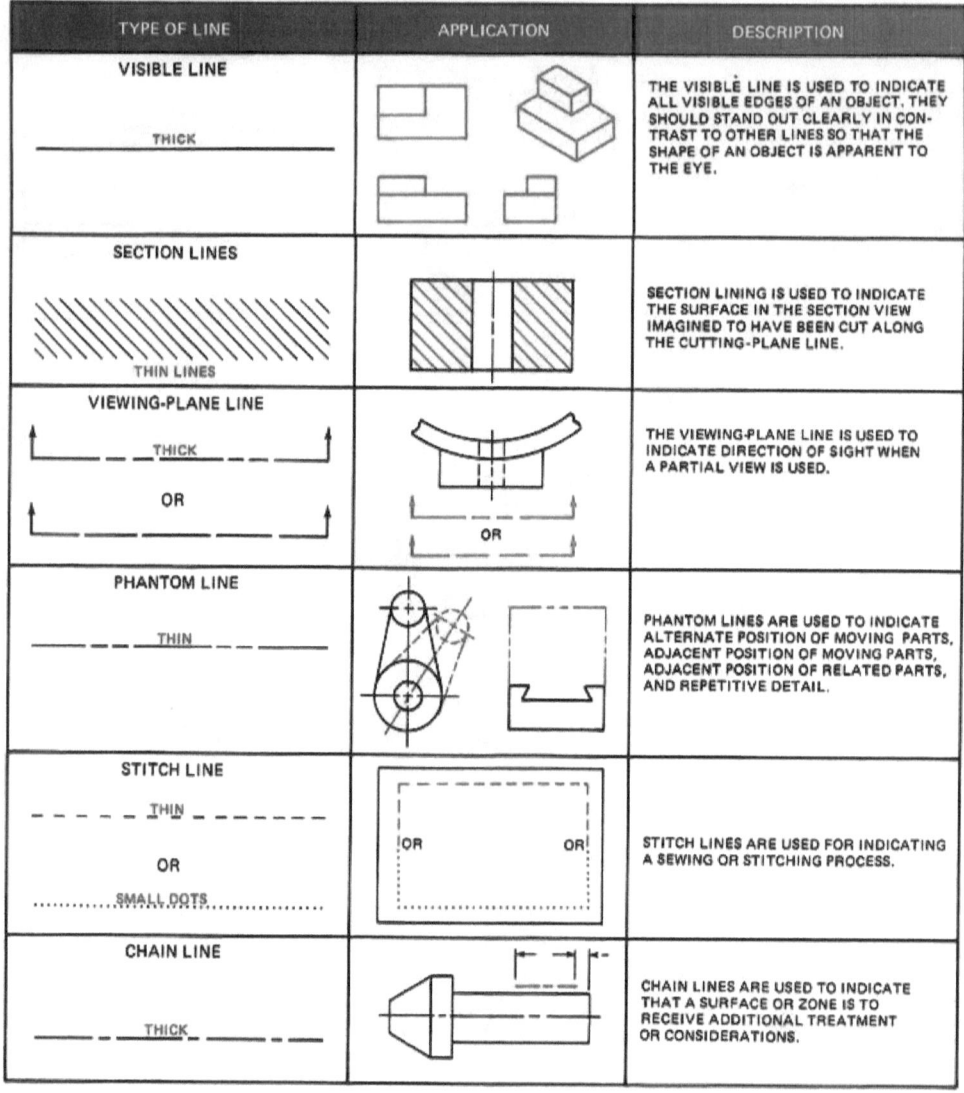

TYPE OF LINE	APPLICATION	DESCRIPTION
VISIBLE LINE THICK		THE VISIBLE LINE IS USED TO INDICATE ALL VISIBLE EDGES OF AN OBJECT. THEY SHOULD STAND OUT CLEARLY IN CONTRAST TO OTHER LINES SO THAT THE SHAPE OF AN OBJECT IS APPARENT TO THE EYE.
SECTION LINES THIN LINES		SECTION LINING IS USED TO INDICATE THE SURFACE IN THE SECTION VIEW IMAGINED TO HAVE BEEN CUT ALONG THE CUTTING-PLANE LINE.
VIEWING-PLANE LINE THICK OR	OR	THE VIEWING-PLANE LINE IS USED TO INDICATE DIRECTION OF SIGHT WHEN A PARTIAL VIEW IS USED.
PHANTOM LINE THIN		PHANTOM LINES ARE USED TO INDICATE ALTERNATE POSITION OF MOVING PARTS, ADJACENT POSITION OF MOVING PARTS, ADJACENT POSITION OF RELATED PARTS, AND REPETITIVE DETAIL.
STITCH LINE THIN OR SMALL DOTS	OR ... OR	STITCH LINES ARE USED FOR INDICATING A SEWING OR STITCHING PROCESS.
CHAIN LINE THICK		CHAIN LINES ARE USED TO INDICATE THAT A SURFACE OR ZONE IS TO RECEIVE ADDITIONAL TREATMENT OR CONSIDERATIONS.

Fig 3.29 Alphabet of Lines used in multiview drawings

3.12.1 Hidden line conventions

In engineering and technical drawing, it is important that hidden features be represented, so that the reader of the drawing can clearly understand the object. Many conventions related to hidden lines have been established over the years. The hidden line conventions that must be followed when creating technical drawings are listed below (see fig. 3.30).

44

i.   There should be no gap when a hidden line intersects a visible line (fig 3.30 A)

ii.   Corners on hidden lines should be joined (fig 3.30 B)

iii. There should be a gap when a hidden line intersects either a visible corner or visible arc (fig 3.30 C)

iv.   Three (hidden) intersecting corners, found in holes that are drilled and that end inside the object (i.e., do not go all the way through the object), should be joined as shown in figure below (fig 3.30 D)

v.   At the bottom of the drilled hole, the lines indicating the tip (Created by the drill, which has a pointed tip) are joined (fig 3.30 E)

vi.   Hidden arcs are started on the center line or the point of tangency (fig 3.30 F)

vii. When a hidden line passes behind a visible line (i.e., does not intersect the visible line), do not put a hidden-line dash on the visible line (fig 3.30 G)

viii. At the point where one hidden line crosses in front of another hidden line (including two hidden features, one closer to the visible view than the other), use a dash for the hidden line in front; that is, if the front hidden line is horizontal, use a horizontal dash at the point of crossing (fig 3.30 H)

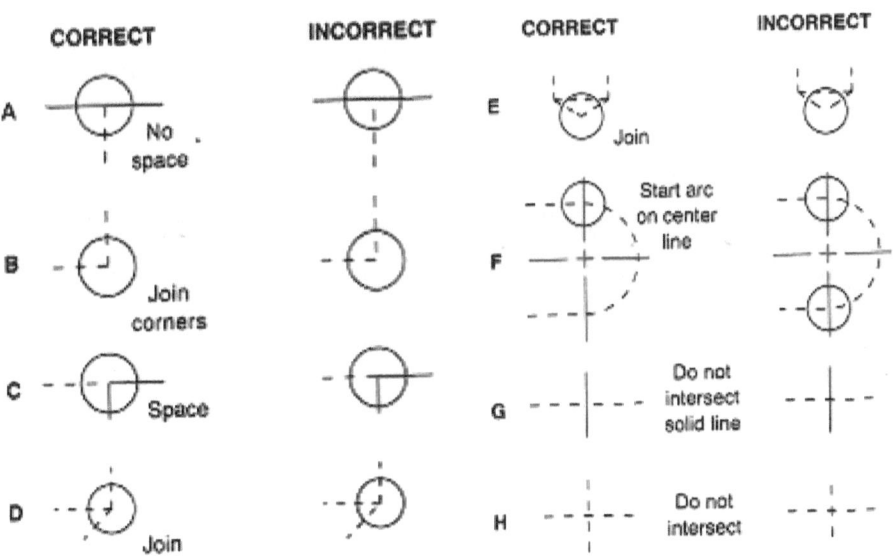

Fig. 3.30 Hidden line conventions

45

3.13   Precedence of lines

In a drawing, full lines, center lines and Hidden lines often coincide and a definite precedence has to be fixed to avoid confusion. The order of precedence of lines is as follows:

1. A full line (Visible line), being most prominent, takes precedence over every other line.
2. Hidden line and Cutting plane line take precedence over center lines.
3. Center line does not have precedence.

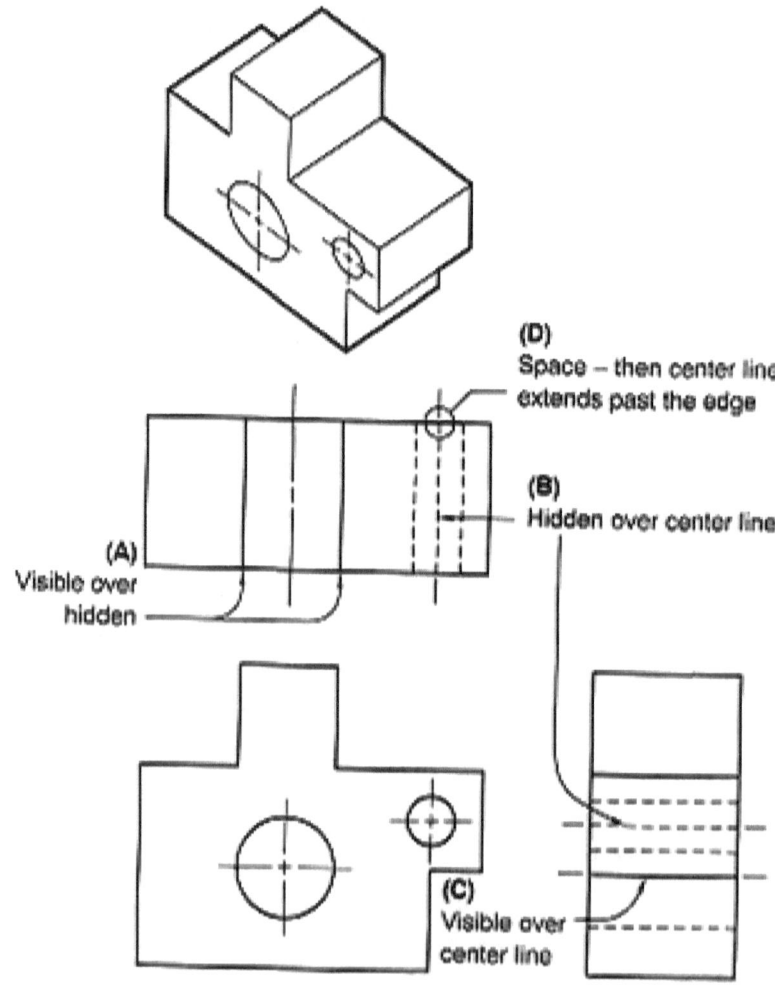

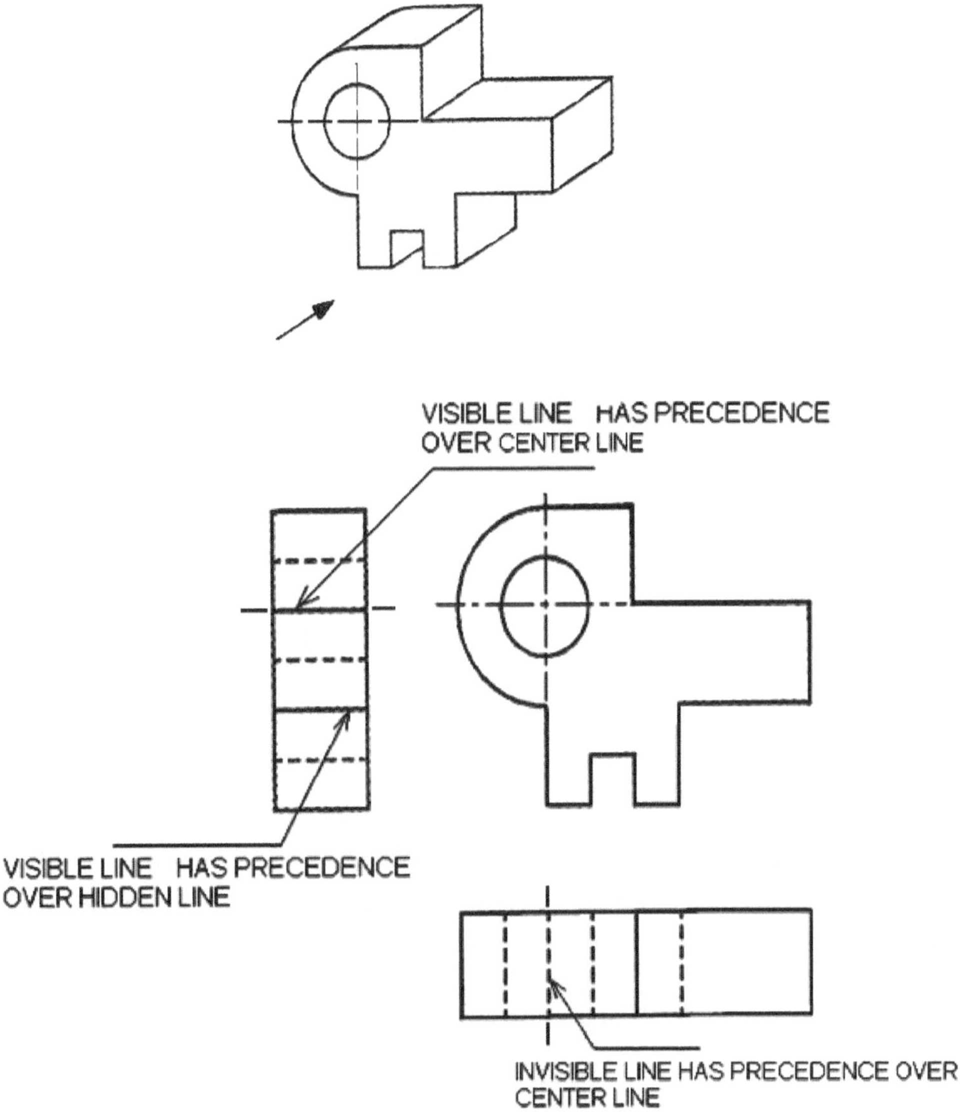

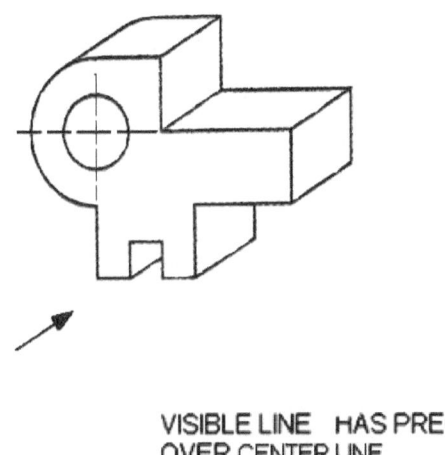

VISIBLE LINE   HAS PRECEDENCE
OVER CENTER LINE

VISIBLE LINE   HAS PRECEDENCE
OVER HIDDEN LINE

INVISIBLE LINE HAS PRECEDENCE OVER
CENTER LINE

Fig. 3.31 Precedence of lines

3.14 Tangent Surfaces

A rounded end (or partial cylinder) is represented as an arc when the line of sight is parallel to the axis of the partial cylinder. No line is drawn at the place where the partial cylinder becomes tangent to another feature, such as the vertical face of the side. When the transition of a rounded end to another feature is not tangent, a line is used at the point of intersection.

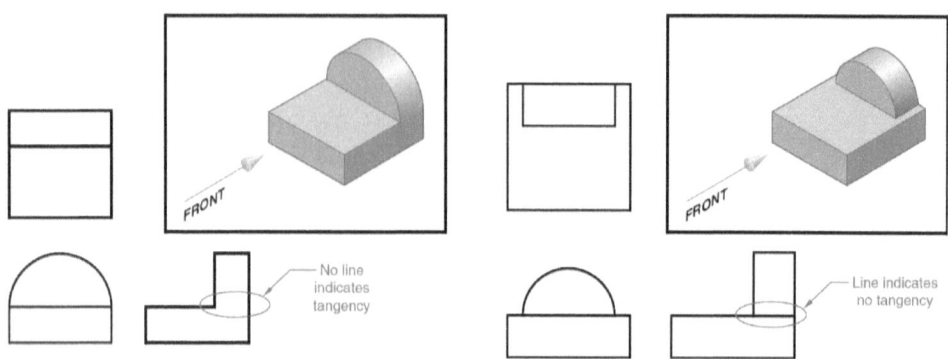

Fig. 3.32 Tangent surfaces

3.15 Fillets, Rounds, and Chamfers

A **fillet** is a rounded interior corner, normally found on cast, forged, or plastic parts. A **round** is a rounded exterior corner, normally found on cast, forged, or plastic parts. A **fillet** or **Round** can indicate that both intersecting surfaces are not machine finished. A fillet or round is shown as a **small arc**. Fillets and Rounds eliminate sharp corners on objects.

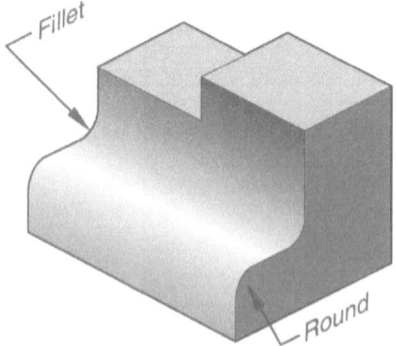

48

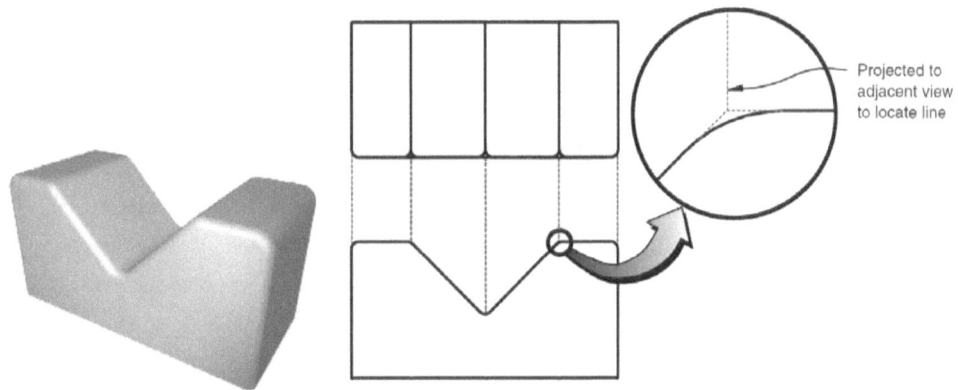

Fig. 3.33 Representing fillet and rounded corners

A **Chamfer** is beveled corner used on the openings of holes and the ends of cylindrical parts, to eliminate sharp corners at the end of cylinders and holes. Chamfers are represented as lines or circles to show the change of plane. Chamfers can be internal or external and specified by a linear and angular dimension.

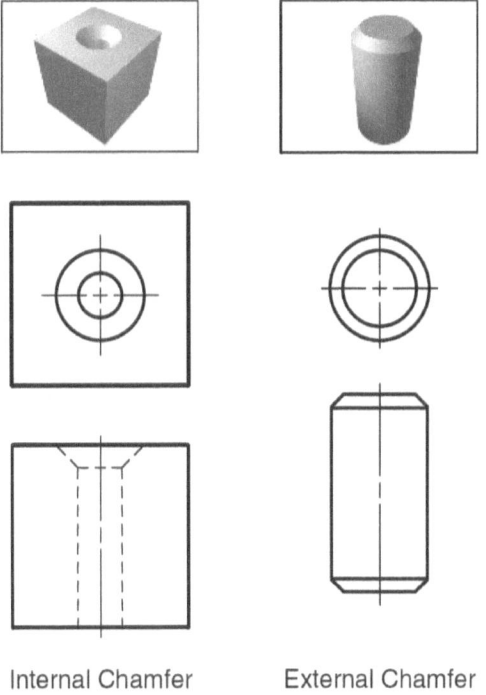

Internal Chamfer   External Chamfer

Fig. 3.34 Examples of internal and external chamfers

3.16   Runouts

A **runout** is a special method of representing filleted surfaces that are tangent to cylinders. A runout is drawn starting at the point of tangency, using a radius equal to that of the filleted surface with a curvature of approximately one-eighth the circumference of a circle. If a very small round intersects a cylindrical surface, the runouts curve away from each other. If a large round intersects a cylindrical surface, the runouts curve toward each other.

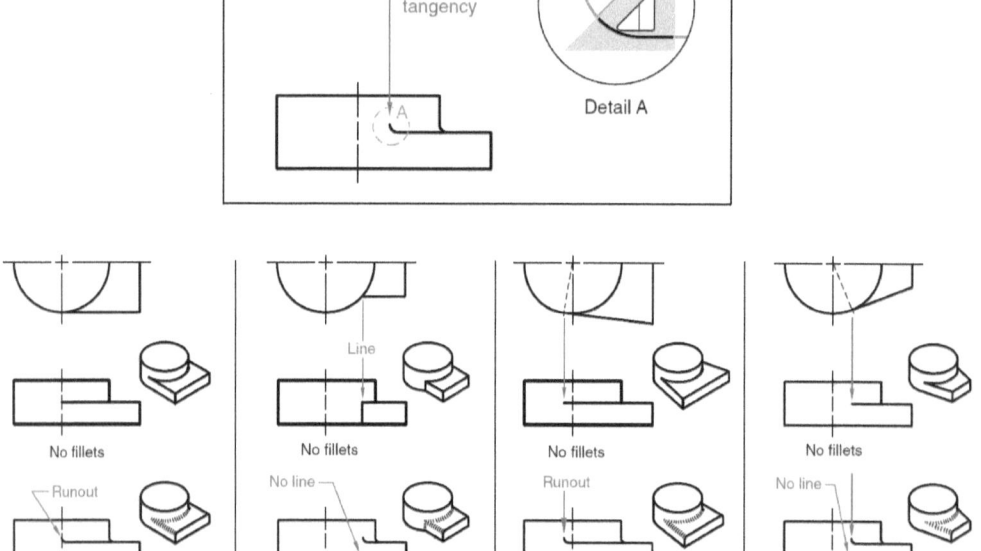

Fig. 3.35 Runouts

Runouts are used to represent corners with fillets that intersect cylinders. Notice the difference in the point of tangency with and without the fillets.

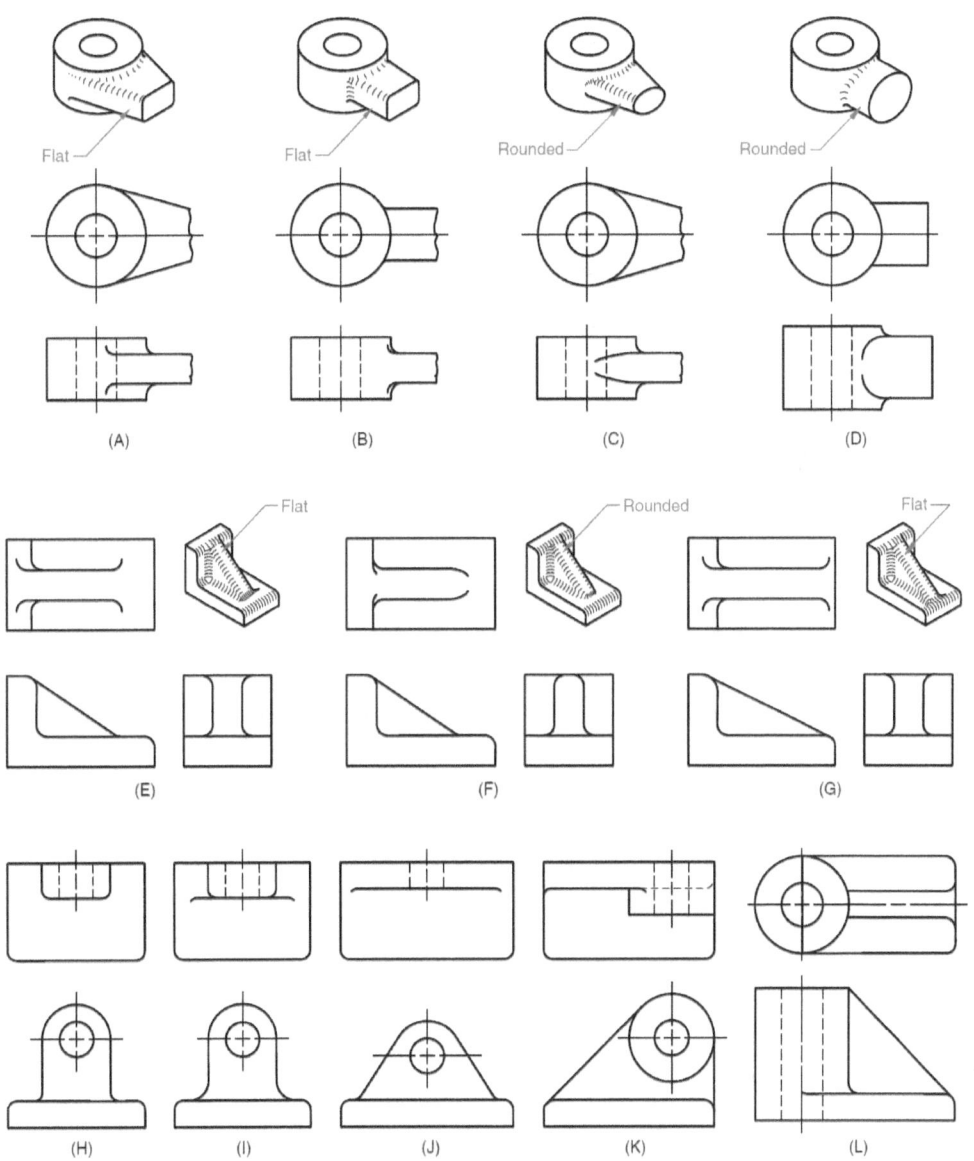

Fig. 3.36 Examples of runouts in multiview drawings

51

3.17   Review Questions

1. Define orthographic projection. How is orthographic projection different from perspective projection? Use a sketch to highlight the differences.
2. Define multiview drawings. Make a simple multiview sketch of an object.
3. Define frontal, horizontal, and profile planes and List the six principal views.
4. Define a normal plane, an inclined plane, and an oblique plane.
5. Sketch multiviews of the objects shown in the pictorials below

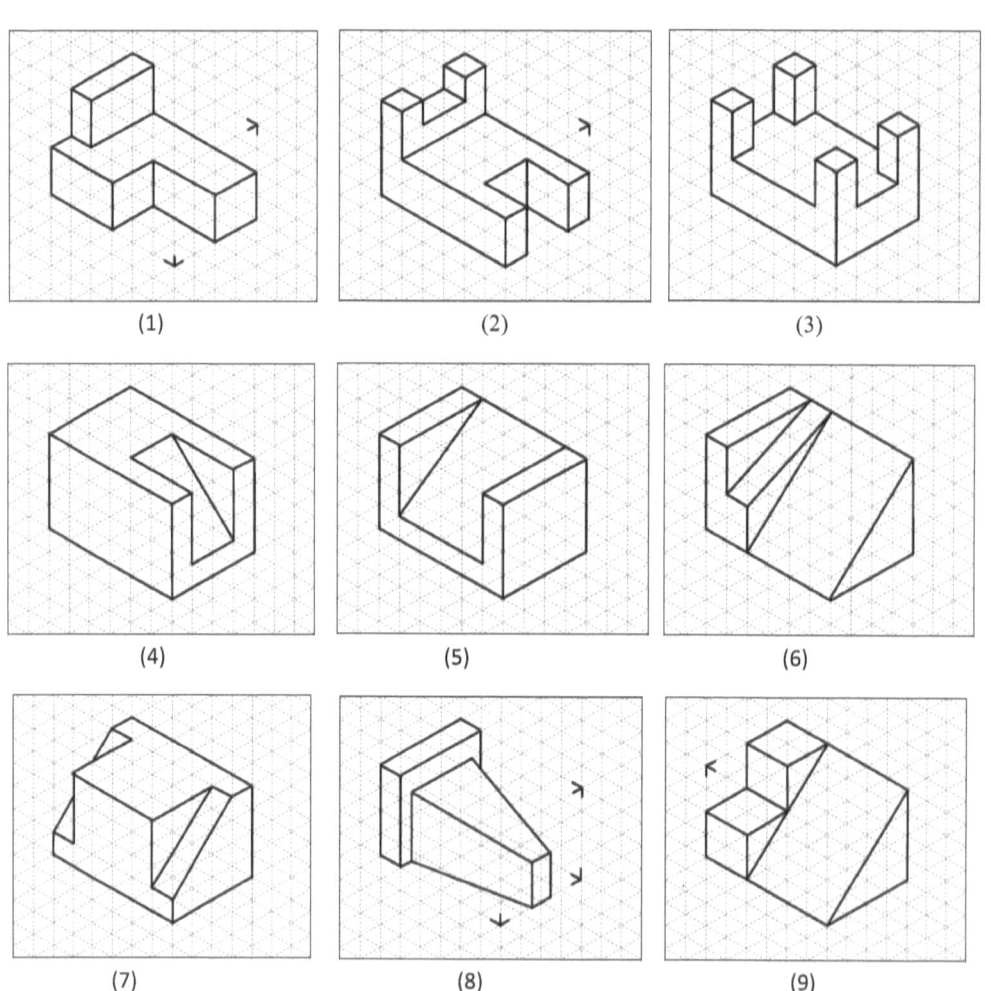

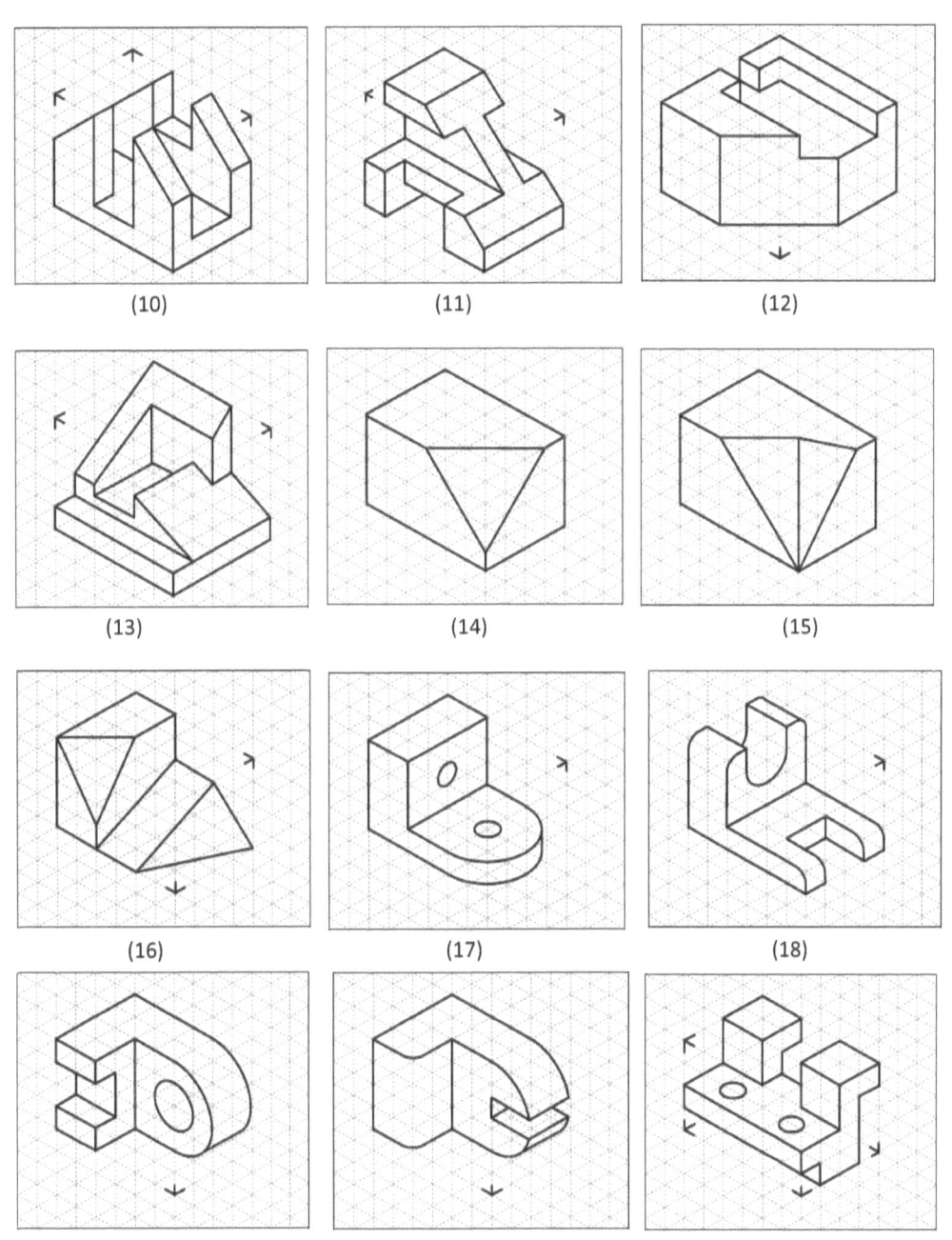

(10)   (11)   (12)

(13)   (14)   (15)

(16)   (17)   (18)

(19)   (20)   (21)

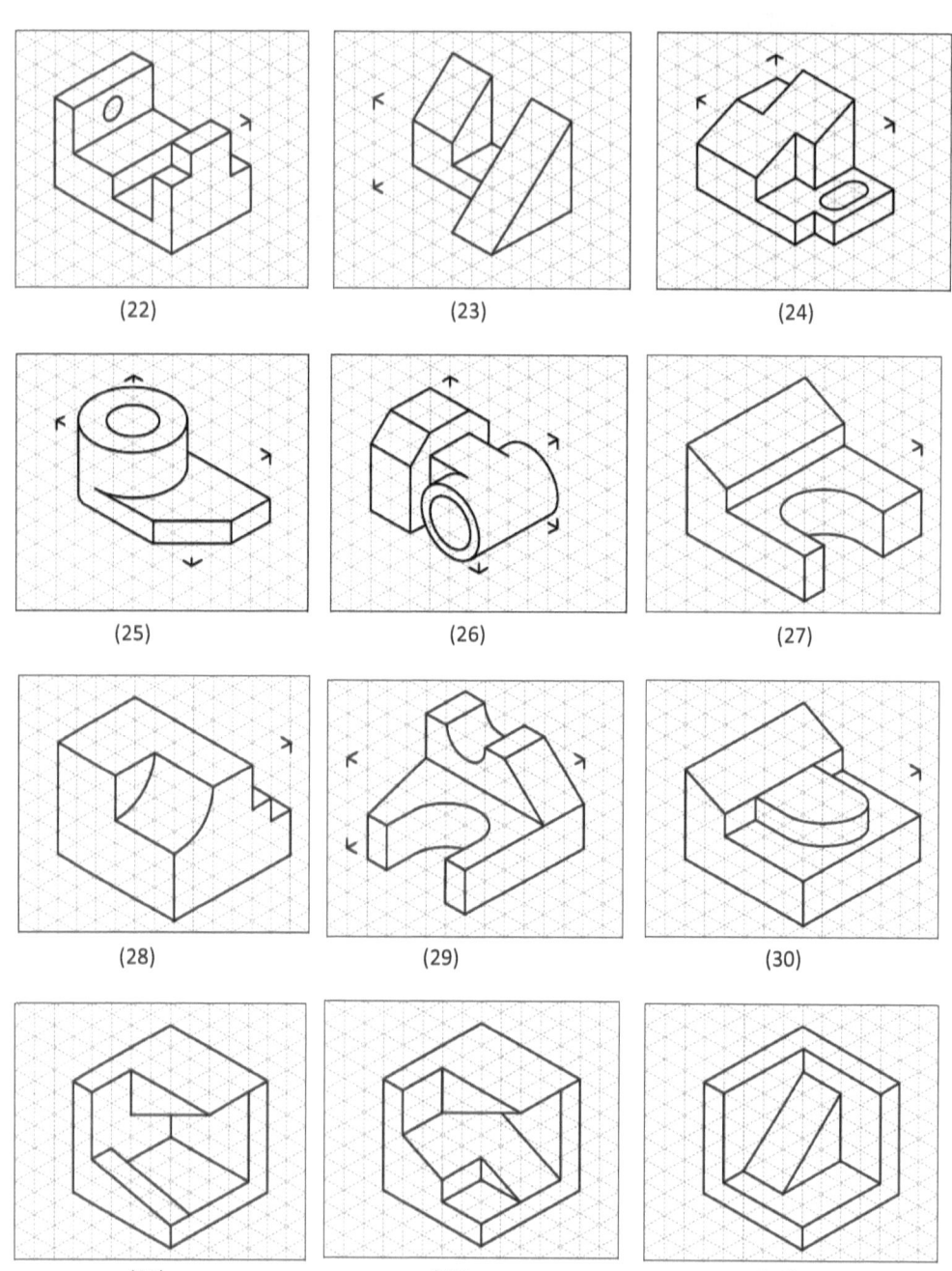

(22)　　　　　　(23)　　　　　　(24)

(25)　　　　　　(26)　　　　　　(27)

(28)　　　　　　(29)　　　　　　(30)

(31)　　　　　　(32)　　　　　　(33)

6. Draw the three principal views **(Front, Top and Right side view)** of the following pictorial drawings using third-angle projection system.

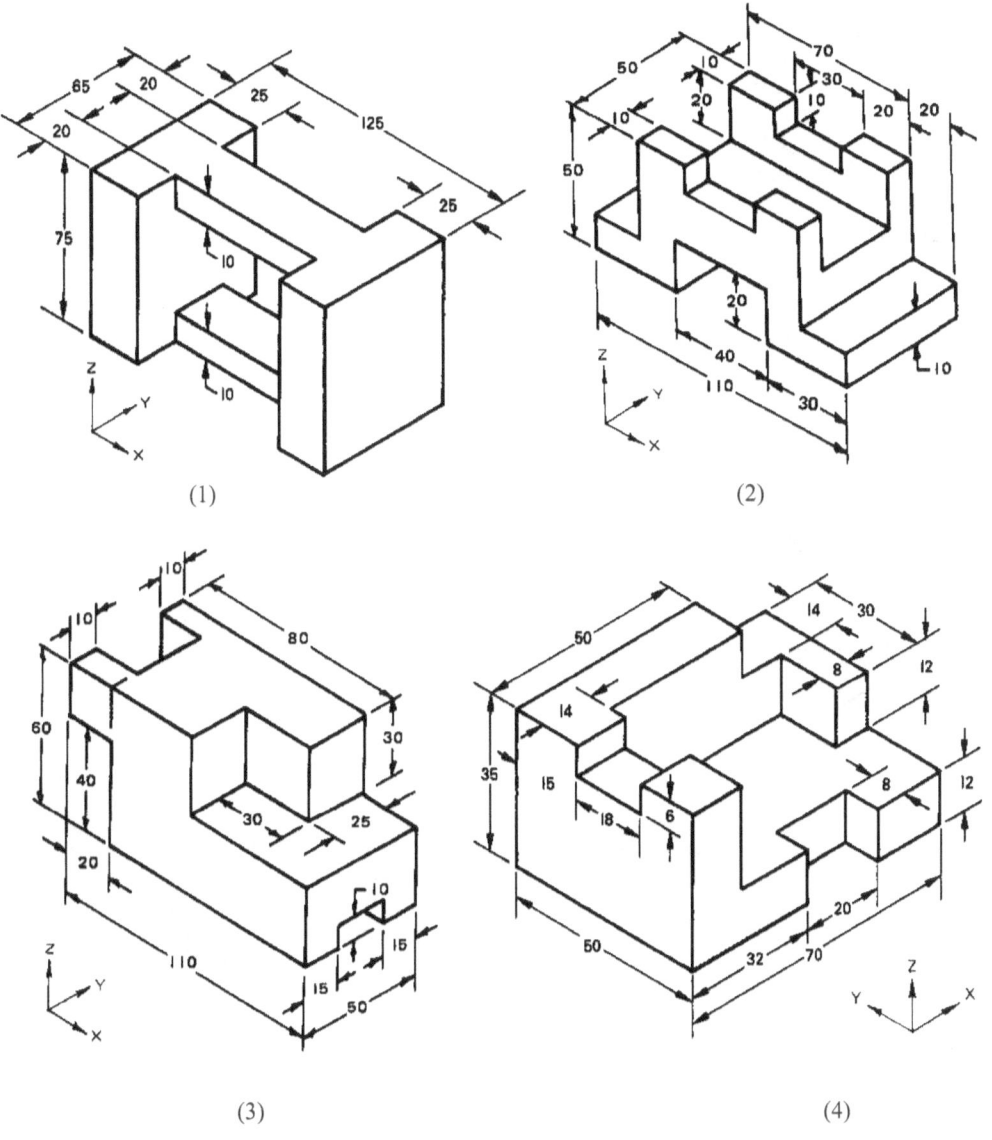

(1)

(2)

(3)

(4)

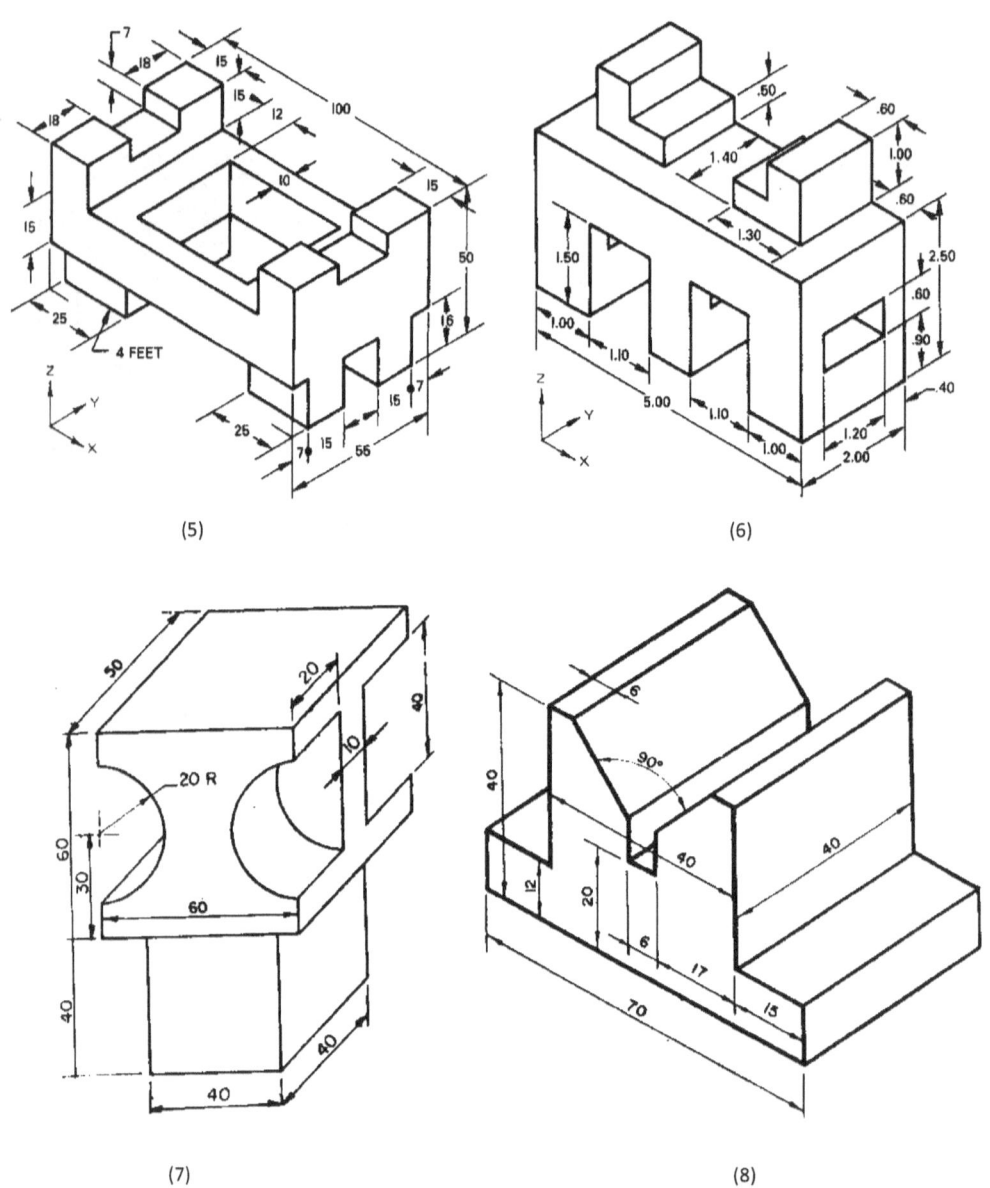

(5)

(6)

(7)

(8)

56

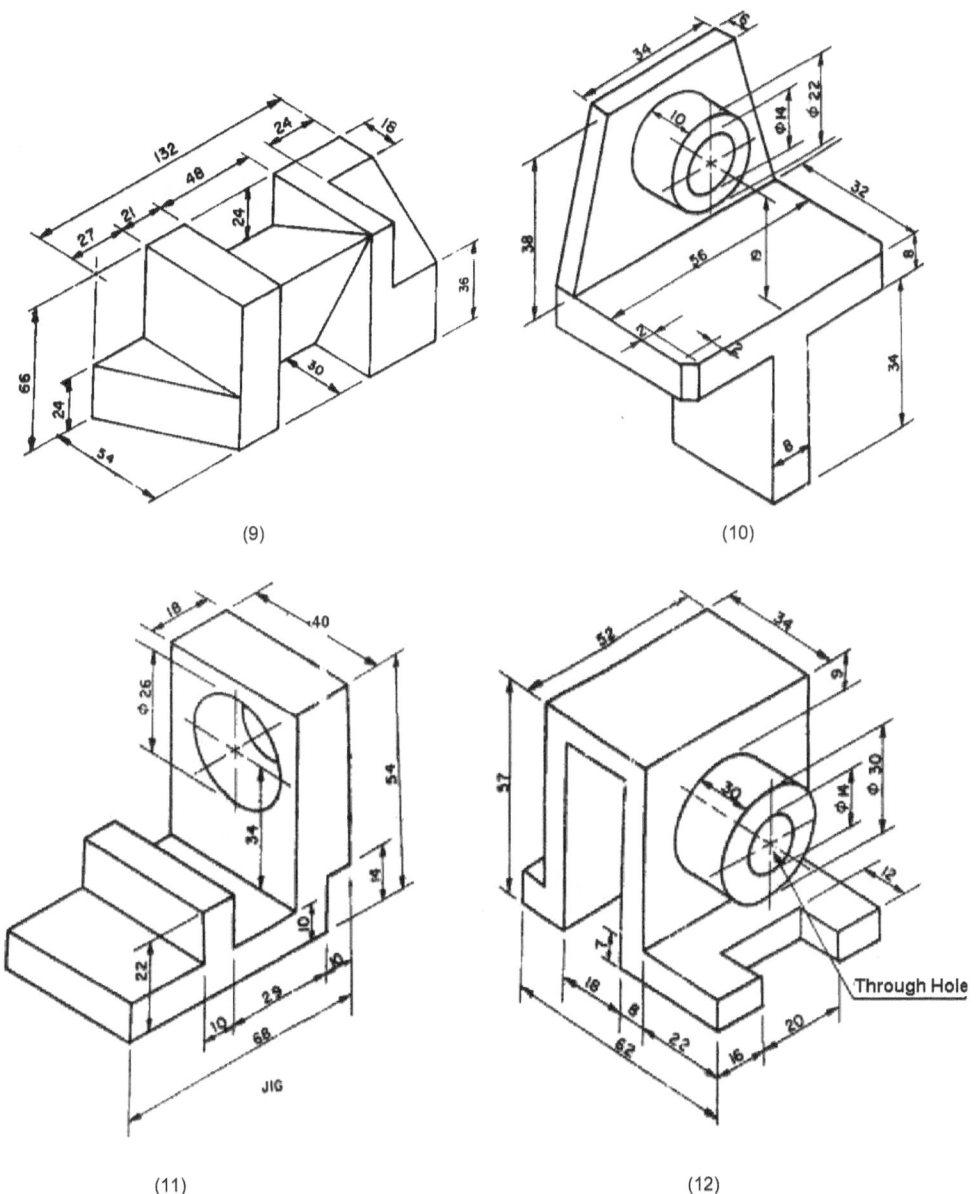

(9)

(10)

(11)

JIG

(12)

Through Hole

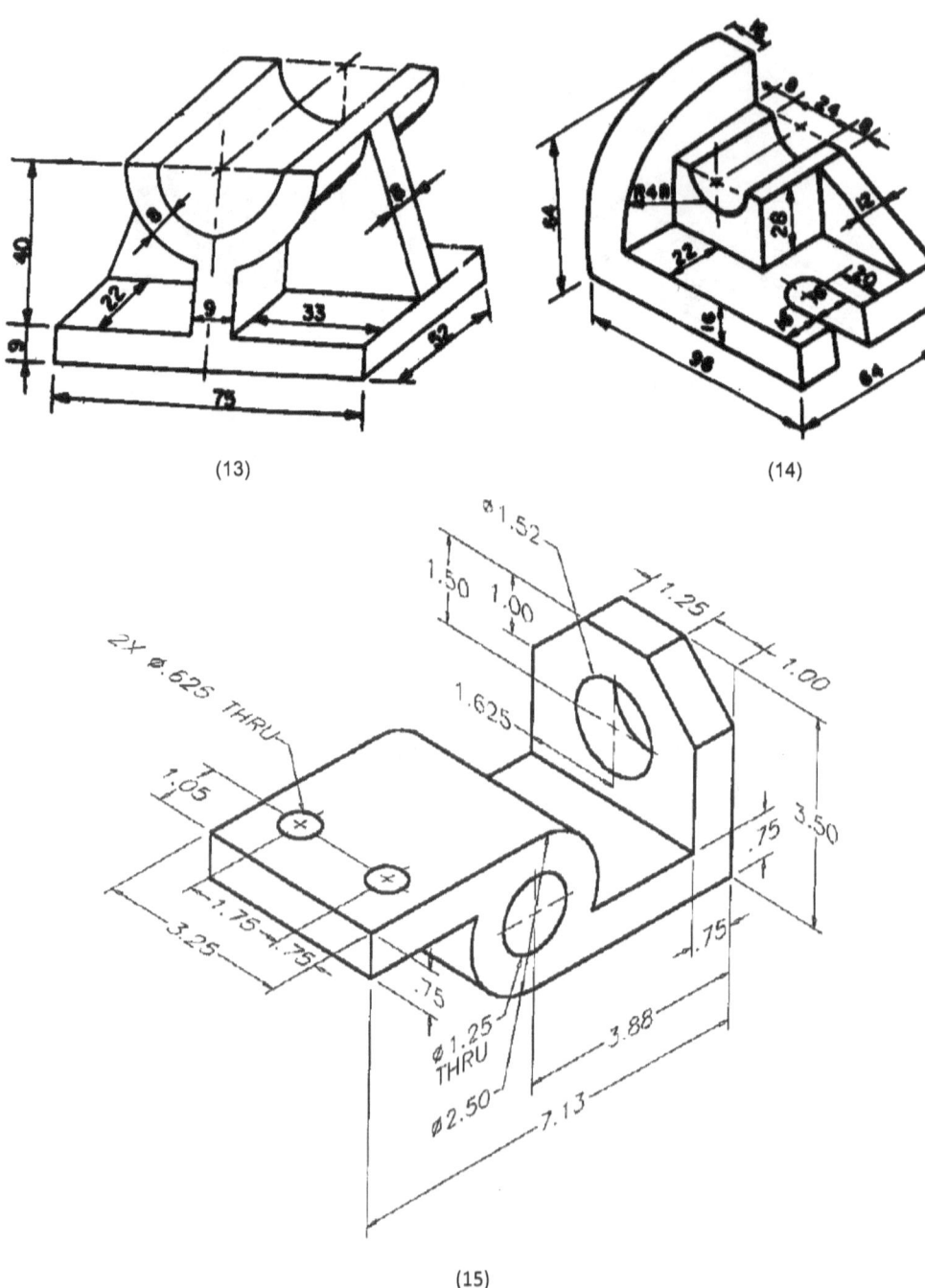

(13)

(14)

(15)

CHAPTER 4

PICTORIAL DRAWINGS

4.1 Introduction

In a multi-view representation of an object, two or more views are used to describe its form and size accurately. However, since each of the views shows only two principal dimensions without any suggestion of depth, such a representation can convey full information only to the professionals who are familiar with graphics language, i.e. Technical drawing. For this reason, multi-view drawings are used mainly by engineers, draftsmen, architects and contractors. To this end, the professionals use conventional picture representations (Pictorial drawings) to communicate with other people who do not possess the required visualization skill to construct an object an object in the mind from its multi-view drawing.

Pictorial drawings show several faces of an object at once as they appear to the observer. They enable a person, without technical training, to visualize an object represented.

In spite of its advantages, a pictorial drawing has its own limitations as compared to a multi- view representation:

> ➤ Pictorial drawing frequently has a distorted and unreal appearance of the object being represented.
> ➤ In many cases, the time required to draw pictorial projection of an object is greater than a multi-view orthographic drawing, and
> ➤ Pictorial drawing is relatively difficult to measure and to give dimensions.

4.2 Types of Pictorial drawings

A pictorial drawing is the representation of a three dimensional object on a two dimensional sheet of paper. The most popular means of a single plane pictorial projections are:

1. Axonometric projection
2. Oblique projection, and
3. Central or Perspective projection.

4.2.1 Axonometric projection

Axonometric projection is a form of orthographic projection with its distinguishing difference that only one plane is used instead of two or more planes to represent an object.

Axonometric projection is a parallel projection technique used to create a pictorial drawing of an object by <u>rotating the object on an axis relative to a projection plane</u>. In axonometric and multiview projections the lines of sight are perpendicular to the plane of projection (thus they are called Orthographic projection). The differences between a multiview drawing and an axonometric drawing are that, in a multiview, only two dimensions of an object are visible in each view and more than one view is required to define the object, whereas, in an axonometric drawing, the object is rotated about an axis to display all the three dimensions, and only one view is required.

Axonometric projection is produced by multiple parallel lines of sight perpendicular to the plane of projection, with the observer at infinity and the object rotated about an axis to produce a pictorial view.

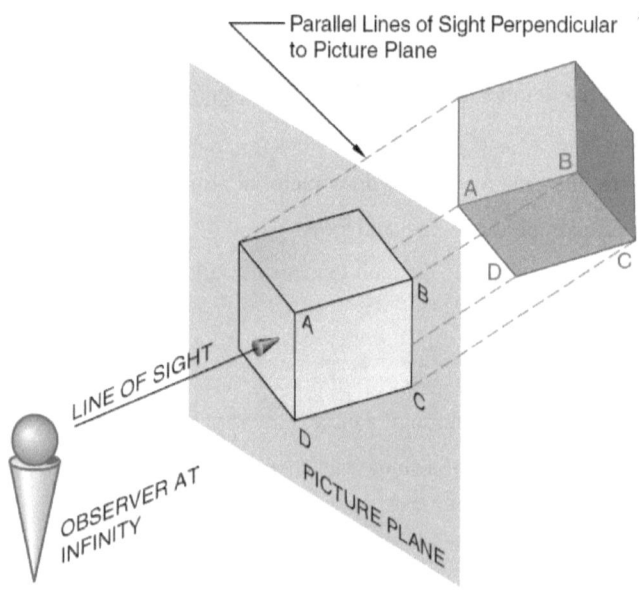

Fig. 4.1 Axonometric Projection

4.2.2 The principles of axonometric projection

To display the axonometric projection of an object, three steps are to be followed in placing the object with respect to the picture plane.

Step I: The object is placed in its customary position, with one principal face parallel to the projection plane.

Step II: It is then rotated from its ordinary position through a proper angle about the vertical axis so that two of its faces will be projected, and

Step III: The object is finally tilted backward through a proper angle so that the third side will also be displayed on the projection plane.

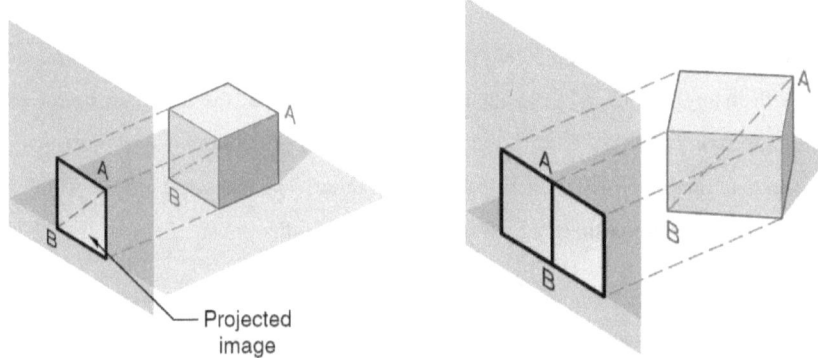

Step 1. Object in normal Orthographic Position   Step 2. Object rotated through an angle

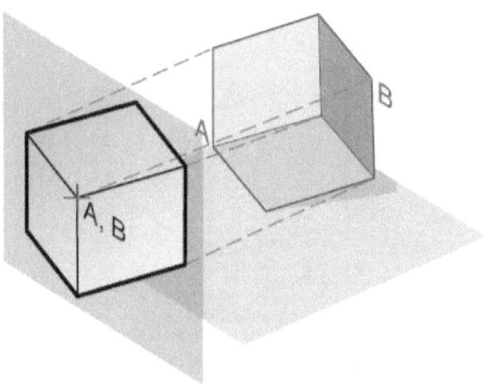

Step 3. Object tilted backward

Fig. 4.2 The principles of axonometric projection

61

4.2.3 Types of Axonometric Drawings

Axonometric drawings are classified by the angles between the lines comprising the axonometric axes. The **axonometric axes** are axes that meet to form the corner of the object that is nearest to the observer.

A. Isometric

Literally the term 'isometric' is a combination of two words 'iso' and 'metric' which means 'equal measures'. It is the simplest type of axonometric projection, because the <u>three principal edges of a cube make equal angles with the projection plane,</u> and hence will be foreshortened equally on the isometric projection. The principal edges are usually referred to as **isometric axes.**

In isometric projection, an object is rotated through 45^0 angle about a vertical axis, and then tilted backward with an angle of 35.25^0 from the top face. In this case, the three isometric axes will make an angle of approximately 35.25^0 with the vertical plane of projection. The projections of these axes will make an angle of 120^0 with respect to each other. The projected length of the principal edges of an object are approximately 81.6% of their true lengths.

Isometric drawings are the quickest and easiest of all the pictorials to draw and <u>the most commonly used</u>. In an isometric drawing the three normal surfaces of a rectangular solid will have equal angles between them (120º).

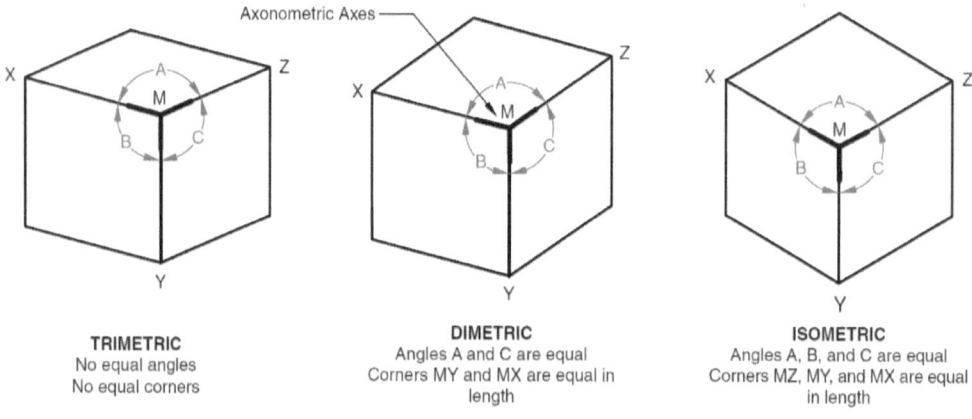

Fig. 4.3 Angles that determine the type of axonometric drawing produced

B. Dimetric

A dimetric projection is an axonometric projection of an object which is placed in such a way that two of its axes make equal angles with the plane of projection, while the third axis makes a different angle with the plane. As a result, the two axes making equal angles are equally foreshortened, while the third edge is foreshortened in a different ratio. In dimetric drawings, two of the normal surfaces will be equally spaced, but the third surface will have an angle of a different number of degrees. Dimetric projection has very little application.

C. Trimetric

A trimetric drawing will have the three normal surfaces of the rectangular solid positioned so none of the three angles have the same number of degrees. Therefore, each of the three principal edges will appear in different length ratios when projected onto a picture plane. The drawing of dimetric and trimetric drawings takes more time because uncommon angles are often used.

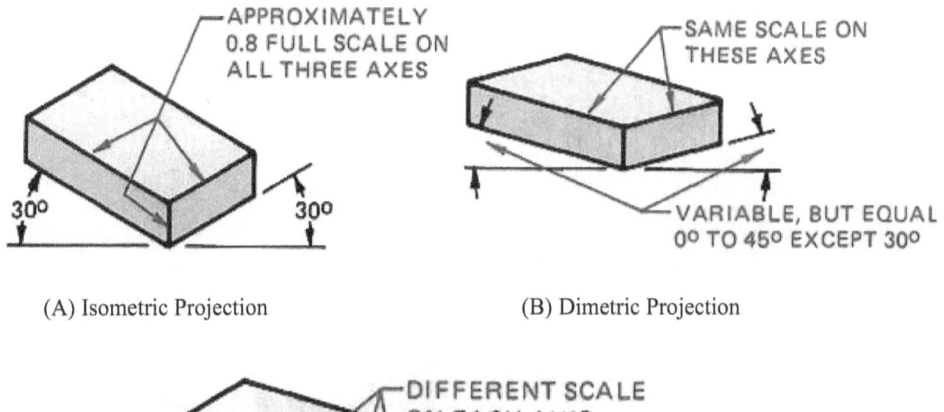

(A) Isometric Projection   (B) Dimetric Projection

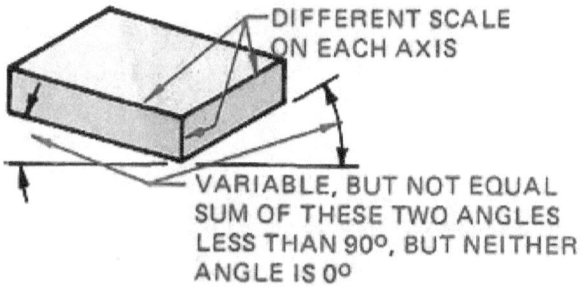

(C) Trimetric Projection

Fig. 4.4 Types of Axonometric projection

63

4.2.4 Isometric Axonometric Projection

An isometric projection is a true representation of the isometric view of an object. An isometric view of an object is created by rotating the object 45 degrees about a vertical axis, then tilting the object (in this case, a cube) forward until the body diagonal (AB) appears as a point in the front view (Fig 4.5). The angle the cube is tilted forward is 35 degrees 16 minutes.

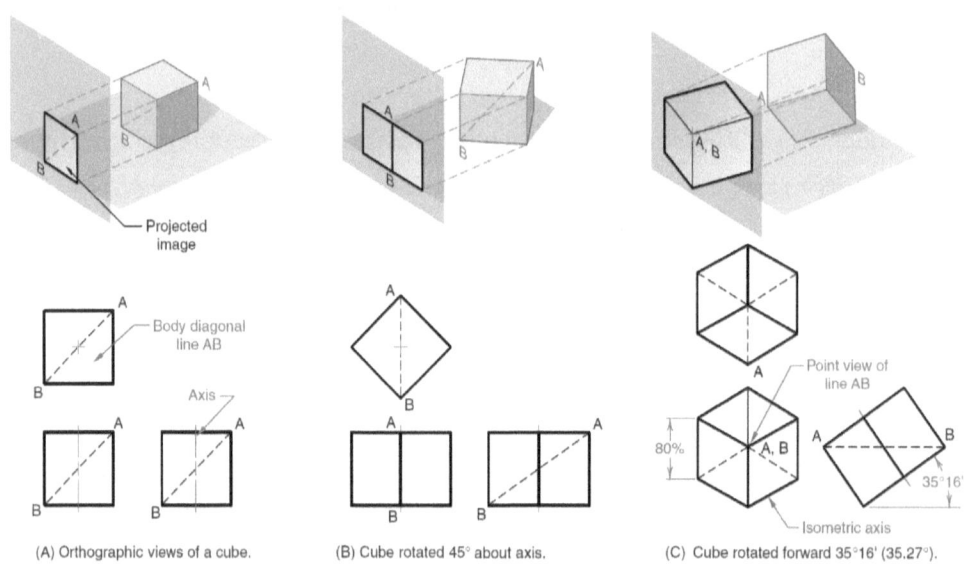

(A) Orthographic views of a cube.   (B) Cube rotated 45° about axis.   (C) Cube rotated forward 35°16' (35.27°).

Fig. 4.5 Theory of isometric projection. The object is rotated 45 degrees about one axis and 35 degrees 16 minutes on another axis

The three axes that meet at A, B form equal angles of 120 degrees and are called the **isometric axes.** Each edge of the cube is parallel to one of the isometric axes. Any line that is parallel to one of the legs of the isometric axis is an **isometric line**. The planes of the cube faces and all planes parallel to them are **isometric planes.**

The forward tilt of the cube causes the edges and planes of the cube to become foreshortened as it is projected onto the picture plane. The lengths of the projected lines are equal to the cosine of 35 degrees 16 minutes, or 0.81647 times the true length. In other words, the projected lengths are approximately 80 percent of the true length.

64

A drawing produced using a scale of 0.816 is called an **isometric projection** and is a true representation of the object. However, if the drawing is produced using a full scale, it is called an **isometric drawing,** which is the same proportion as an isometric projection but is <u>larger by a factor of 1.23 to 1</u>. Isometric drawings are almost always preferred over isometric projection for technical drawings because they are easier to produce.

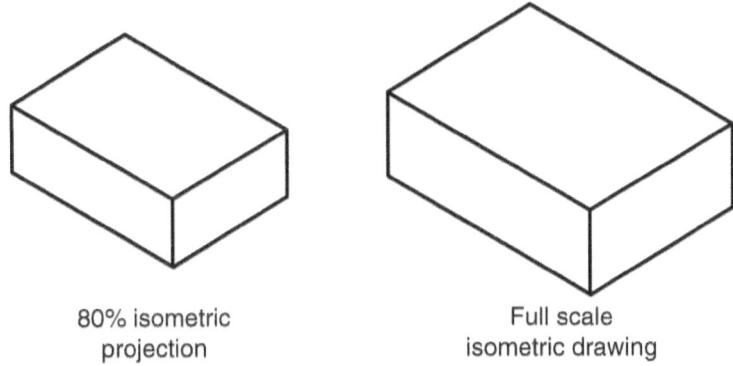

80% isometric
projection

Full scale
isometric drawing

Fig. 4.6 The different scales of an Isometric projection and an isometric drawing

A true isometric projection shortens all the three dimensions by approximately 80%. The more common isometric drawing is created at full scale.

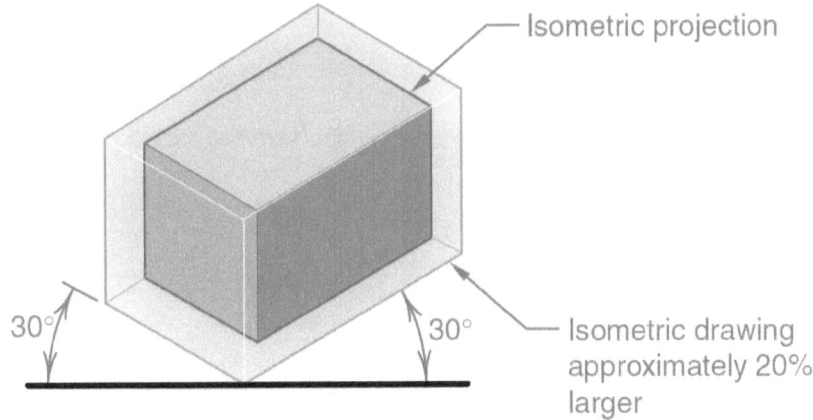

Fig. 4.7 Size comparison of isometric drawing and true isometric projection

4.3 Isometric Axonometric Drawings

An isometric drawing is an axonometric pictorial drawing for which the angle between each axis equals 120 degrees and the scale used is full scale. Isometric axes can be positioned in a number of different ways to create different views of the same object.

➤ **Regular isometric:** The viewpoint is looking down on the top of the object. In a regular isometric, the axes at 30 degrees to the horizontal are drawn upward from the horizontal. The regular isometric is the most common type of isometric drawing.

➤ **Reversed axis isometric:** The view point is looking up on the bottom of the object, and the 30-degree axes are drawn downward from the horizontal.

➤ **Long axis isometric:** The view point is **looking** from the right or left of the object, and one axis drawn at 60 degrees to the horizontal. This is applicable for long slender object, which has a considerable width as compared to its depth and height, the long axis may be placed horizontal.

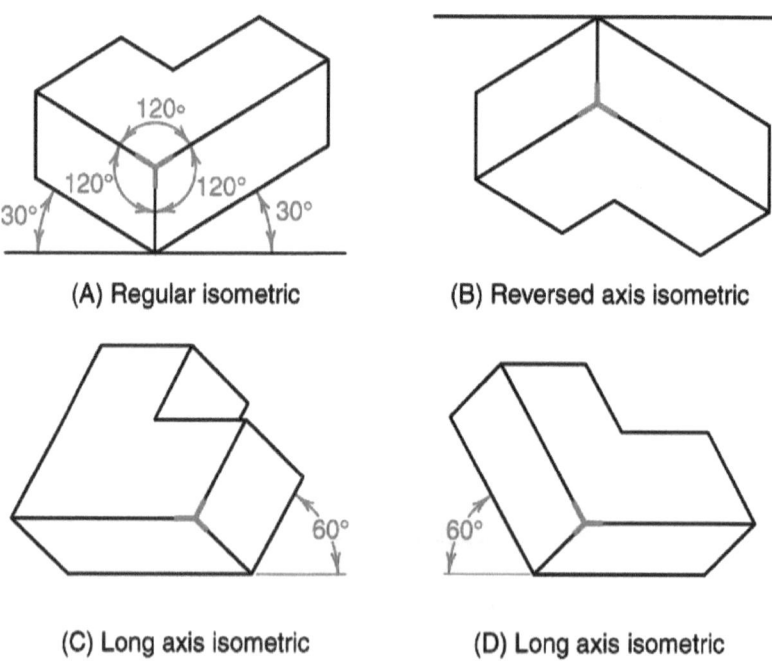

Fig. 4.8 Positions of isometric axes and their effect on the view created

In an isometric drawing, true length distances can only be measured along isometric lines, that is, lines that run parallel to any of the isometric axes. Any line that does not run parallel to an isometric axis is called a **nonisometric line**. Nonisometric lines include inclined and oblique lines and cannot be measured directly. Instead, they must be created by locating two endpoints.

Isometric planes are parallel to the isometric surfaces formed by any two adjacent isometric axes. Planes that are not parallel to any isometric plane are called **nonisometric planes.**

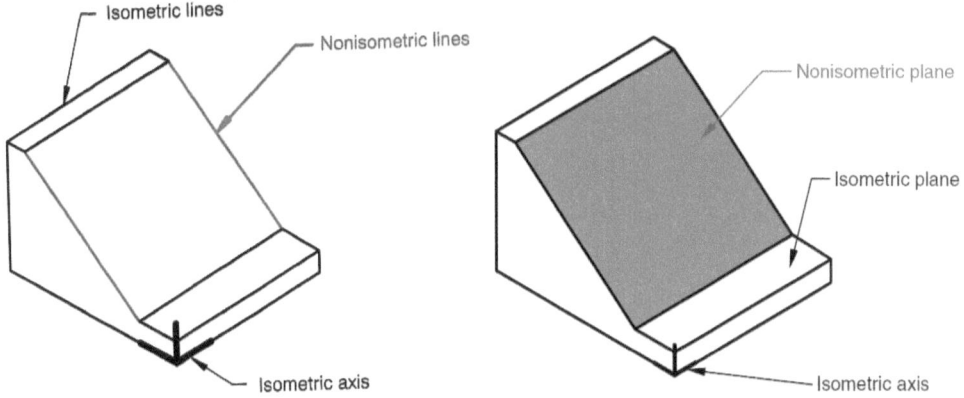

Fig. 4.9 Isometric and nonisometric lines and planes

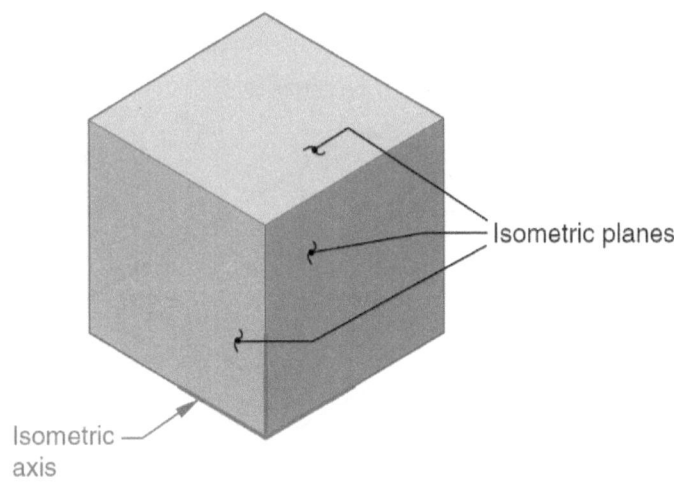

Fig. 4.10 Isometric planes relative to isometric axes

4.3.1 Standards for Hidden Lines, center lines, and Dimensions

In isometric drawings **hidden lines** are omitted unless absolutely necessary to completely describe the object. Most isometric drawings will not have hidden lines. To avoid using hidden lines, choose the most descriptive viewpoint. However, if an isometric view point cannot be found that clearly depicts all the major features, hidden lines may be used.

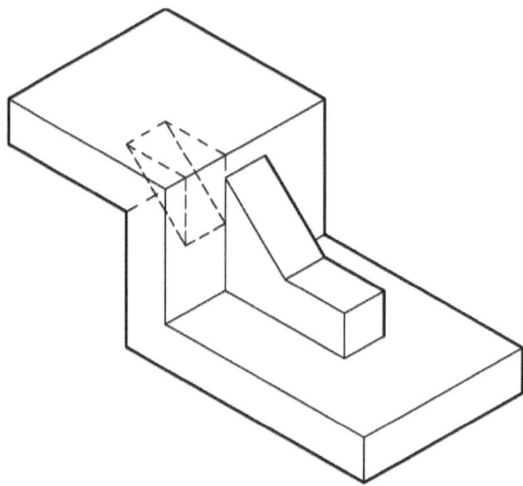

Fig. 4.11 An isometric drawing with hidden lines for details not otherwise clearly shown

In isometric drawings center lines are drawn only for showing symmetry or dimensioning. Normally, center lines are not shown because many isometric drawings are used to communicate to nontechnical people and not for engineering purposes.

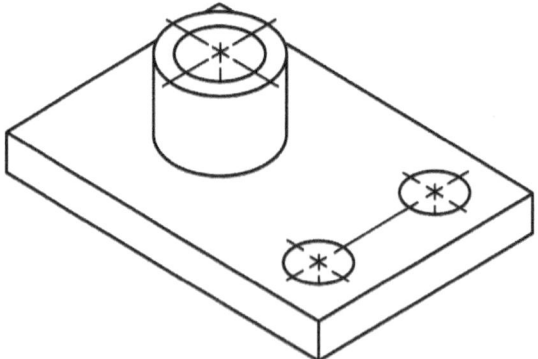

Fig. 4.12 Center lines used for dimensioning

Dimensioned isometric drawings used for production purposes must be ANSI standard, with dimension and extension lines and lines to be dimensioned lying in the same plane. All dimensions and notes should be unidirectional, reading from the bottom of the drawing upward, and should be located outside the view whenever possible. Dimensioned drawings used for illustration purposes (non-technical) may use the aligned method.

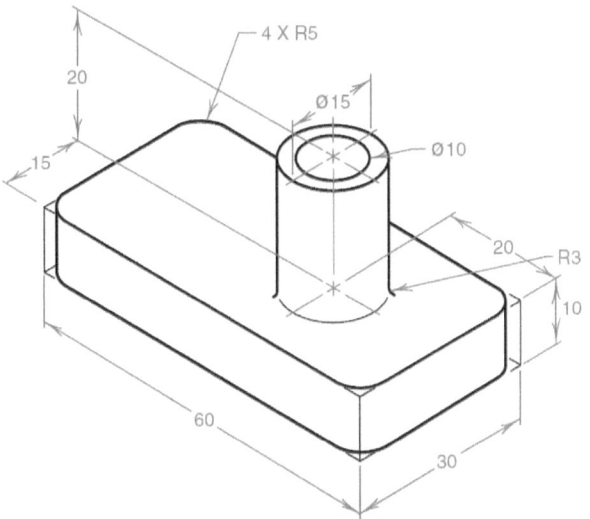

Fig. 4.13 ANSI Standard Unidirectional Isometric Dimensioning

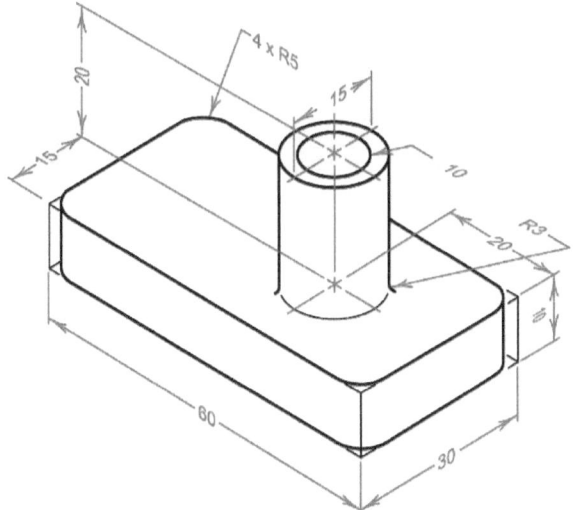

Fig. 4.14 Aligned Isometric Dimensioning Used for Illustrations

69

4.3.2   The Boxing-In Method for Creating Isometric Drawings

The four basic steps for creating an isometric drawing are as follows:

1. Determine the isometric viewpoint that clearly depicts the features of the object, and then draw the isometric axes that will produce that viewpoint.
2. Construct isometric planes, using the overall width (W), height (H), and depth
3. of the object, such that the object will be totally enclosed in a box.
4. Locate details on the isometric planes.
5. Darken all visible lines, and eliminate hidden lines unless absolutely necessary to describe the object.

Figure 4.15 shows a dimensioned multiview drawing and the steps used to create an isometric drawing of the object, using the boxing-in method.

Step 1. Determine the desired view of the object, and then draw the isometric axes. For this example, it is determined that the object will be viewed from above (regular isometric).

Step 2. Construct the front isometric plane using the W and H dimensions. Width dimensions are drawn along 30-degree lines from the horizontal. Height dimensions are drawn as vertical lines.

Step 3. Construct the top isometric plane using the Wand D dimensions. Both the Wand D dimensions are drawn along 30-degree lines from the horizontal.

Step 4. Construct the right side isometric plane using the D and H dimensions. Depth dimensions are drawn along 30-degree lines and height dimensions are drawn as vertical lines.

Step 5. Transfer some distances for the various features from the multiview drawing to the isometric lines that make up the isometric rectangle. For example, distance A is measured in the multiview drawing, and then transferred to a width line in the front plane of the isometric rectangle. Begin drawing details of the block by drawing isometric lines between the points transferred from the multiview drawing. For example, a notch is taken out of the block by locating its position on the front and top planes of the isometric box.

Step 6. Transfer the remaining features from the multiview drawing to the isometric drawing. Block in the details by connecting the endpoints of the measurements taken from the *multiview* drawing.

***Step* 7.** Darken all visible lines, and erase or lighten the construction lines to complete the isometric drawing of the object

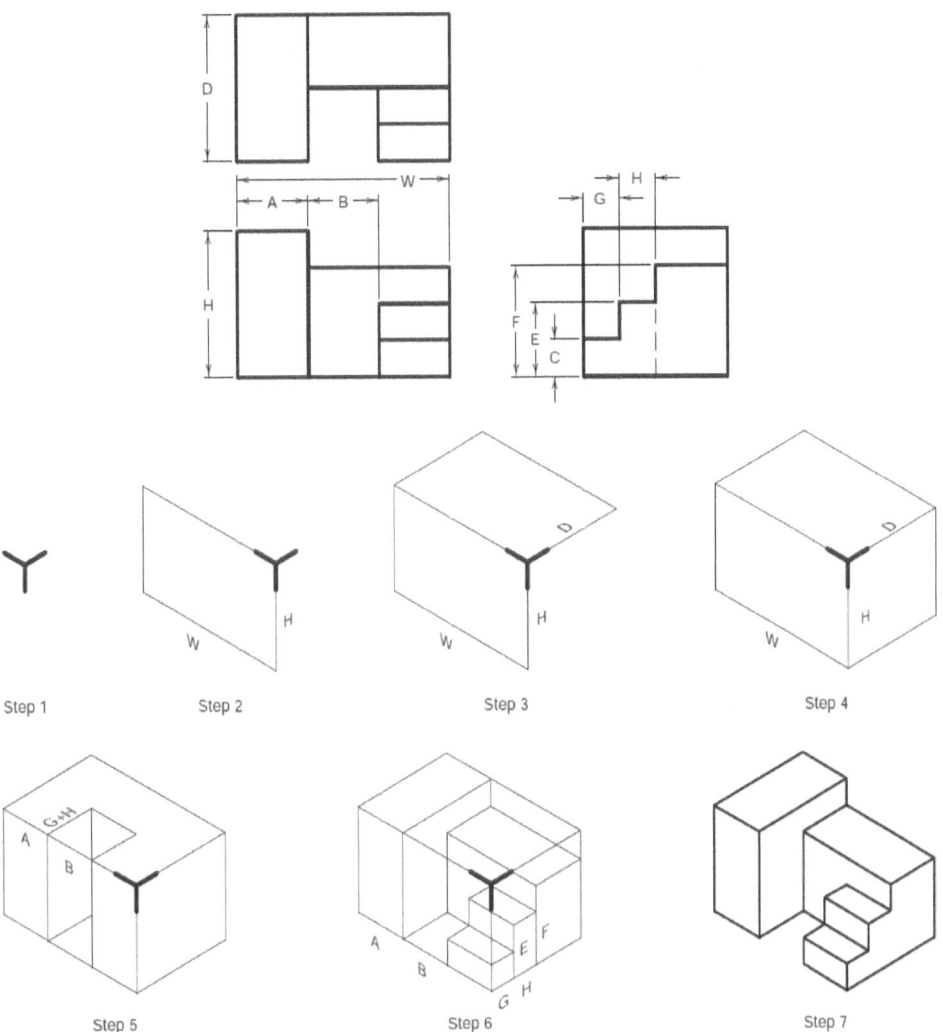

Fig. 4.15 Constructing an Isometric Drawing using the Boxing-In Method

4.3.3 Nonisometric Lines

Normally, nonisometric lines will be the edges of inclined or oblique planes of an object as represented in a multiview drawing. It is not possible to measure the length or angle of an inclined or oblique line in a multiview drawing and then use that measurement to draw the line in an isometric drawing. Instead, nonisometric lines must be drawn by locating the two endpoints, then connecting the endpoints with a line. The process used is called offset measurement, which is a method of locating one point by projecting another point. To create an isometric drawing of an object with nonisometric lines, use the following procedure.

Step 1. Determine the desired view of the object. Then draw the isometric axes.

Step 2. Construct the front isometric plane using the Wand H dimensions.

Step 3. Construct the top isometric plane using the Wand D dimensions.

Step 4. Construct the right side isometric plane using the D and H dimensions

Step 5. Transfer the distances for C and A from the multiview drawing to the top and right side isometric rectangles. Draw line 1-2 across the top face of the isometric box. Draw an isometric construction line from the endpoint marked for distance C. This in effect projects distance C along the width of the box.

Step 6. Along these isometric construction lines, mark off the distance 8, thus locating points 4 and 3. Connect points 4-3.

Step 7. Connect points 1-4 and 2-3 to draw the nonisometric lines.

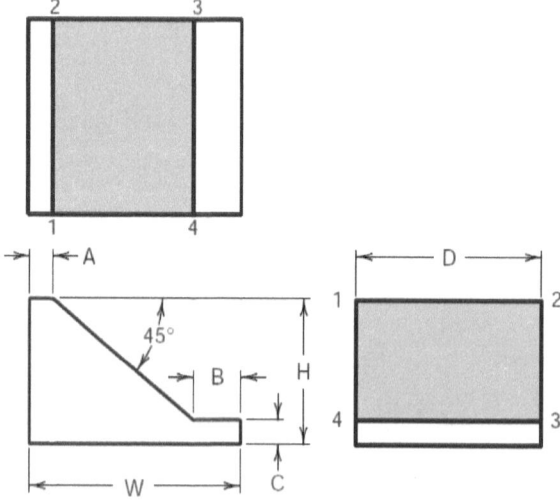

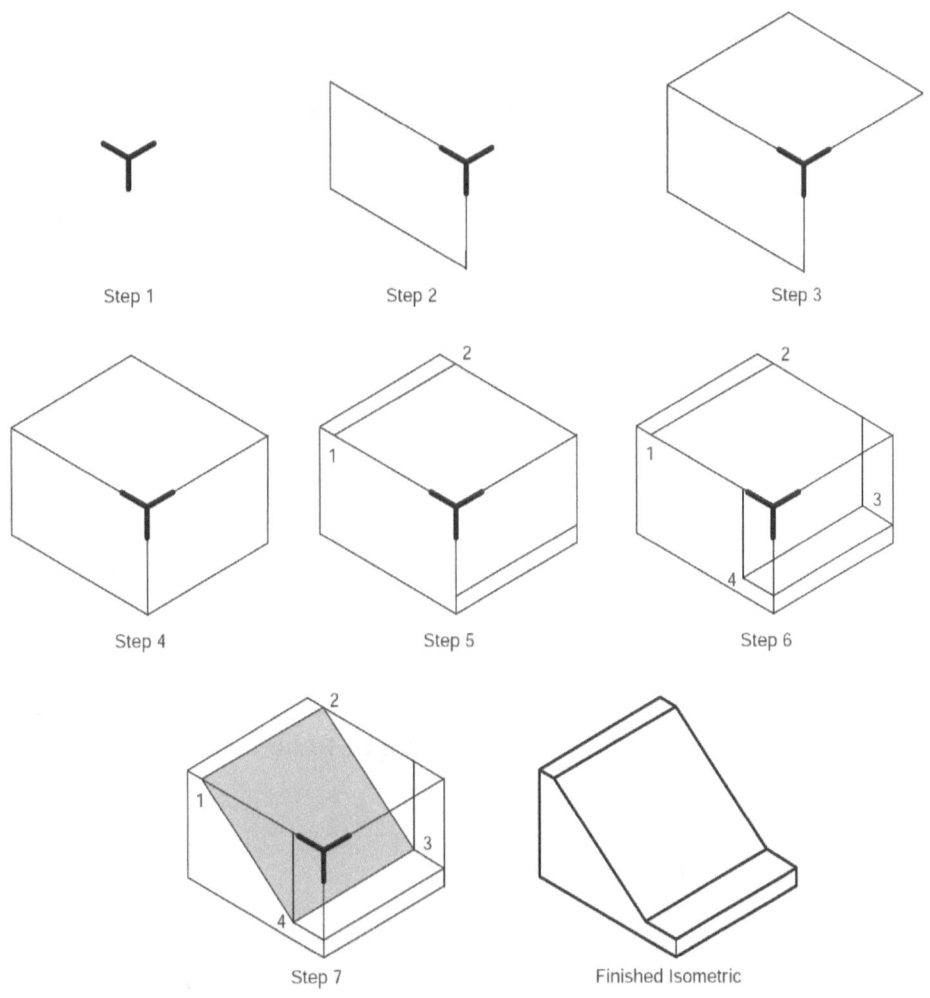

Fig. 4.16 Constructing an isometric drawing having nonisometric lines.

4.3.4 Oblique Planes in isometric views

The initial steps used to create an isometric drawing of an object with an oblique plane are the same as the steps used to create any isometric view. The sides of the oblique plane will be nonisometric lines, which means that their endpoints will be located by projections along isometric lines. After each endpoint is located, the oblique plane is drawn by connecting the endpoints. The following steps describe how to create an isometric view of the multiview drawing with an oblique plane shown in Figure 4.7 below.

73

***Step* 1.** Determine the desired view of the object, and then draw the isometric axes.

***Step* 2.** Construct the front isometric plane using the width and height dimensions.

***Step* 3.** Construct the top isometric plane using the width and depth dimensions.

***Step* 4.** Construct the right side isometric plane using the depth and height dimensions.

***Step* 5.** Locate the slot across the top plane by measuring distances E, F, and G along

   isometric lines.

***Step* 6.** Locate the endpoints of the oblique plane in the top plane by locating distances A, 8, C, and D along the lines created for the slot in Step 5. Label the endpoint of line A as 5, line 8 as 1, line C as 4, and line D as 7. Locate distance H along the vertical isometric line in the front plane of the isometric box and label the endpoint 6. Then locate distance I along the isometric line in the profile isometric plane and label the endpoint 8. Connect endpoints 5-7 and end points 6-8. Connect points 5-6 and 7-8.

***Step* 7.** Draw a line from point 4 parallel to line 7-8. This new line should intersect at point 3. Locate point 2 by drawing a line from point 3 parallel to line 4-7 and equal in length to the distance between points 1 and 4. Draw a line from point 1 parallel to line 5-6. This new line should intersect point 2.

***Step* 8.** Darken lines 4-3, 3-2, and 2-1 to complete the isometric view of the object.

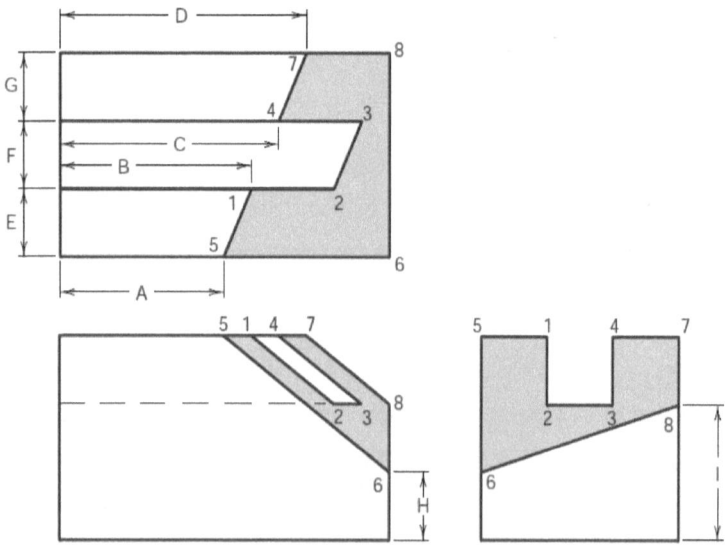

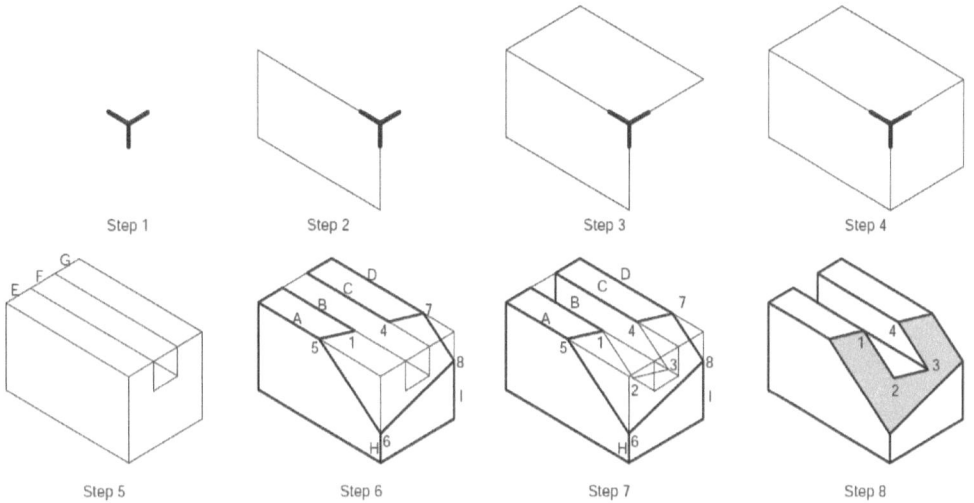

Fig. 4.17 Constructing an Isometric Drawing Having an Oblique Surface.

4.3.5 Angles in isometric views

Angles can only be drawn true size when they are perpendicular to the line of sight. In isometric drawings, this is usually not possible; therefore, angles cannot be measured directly in isometric drawings. To draw an angle in an isometric drawing, locate the endpoints of the lines that form the angle, and draw the lines between the endpoints. The following steps demonstrate the method for drawing an isometric angle.

> ***Step 1.*** Determine the desired view of the object. Then draw the isometric axes.
>
> ***Step 2.*** Construct the front isometric plane using the W and H dimensions. Construct the top isometric plane using the W and D dimensions. Construct the right side isometric plane using the D and H dimensions.
>
> ***Step 3.*** Determine dimensions X and Y from the front view and transfer them to the front face of the isometric drawing. Project distance X along an isometric line parallel to the W line. Project distance Y along an isometric line parallel to the H line. Point Z will be located where the projectors for X and Y intersect.

Step 4. Draw lines from point Z to the upper corners of the front face. Project point Z to the back plane of the box on an isometric line parallel and equal in length to the D line. Draw lines to the upper corner of the back plane to complete the isometric drawing of the object.

Notice that the 45-degree angles do not measure 45 degrees in the isometric view. This is an example of why no angular measures are taken from a multiview to construct an isometric drawing.

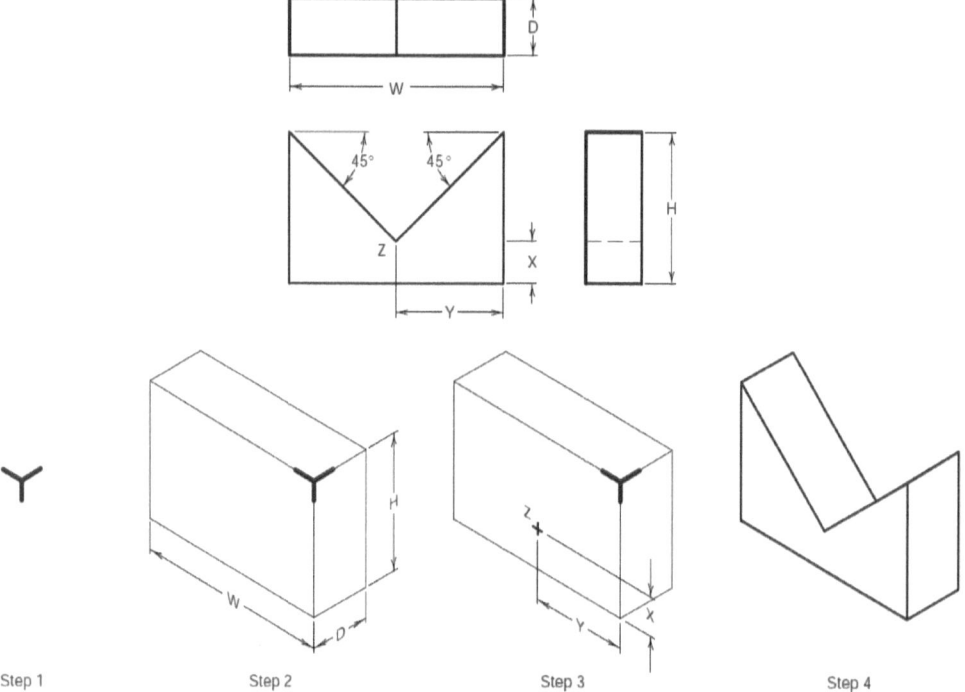

Fig. 4.18 Constructing Angles in an Isometric Drawing

4.3.6   Constructing Irregular Curves in an Isometric Drawing

Fig. 4.19 and the following steps describe how to create an irregular isometric curve.

Step 1. On the front view of the multiview drawing of the curve, construct parallel lines, and label the points 1-12. Project these lines into the top view until they intersect the curve. Label these points of intersection 13-18, as shown in Fig. 4.19. Draw horizontal lines through each point of intersection, to create a grid of lines.

Step 2. Use the W, H, and D dimensions from the multiview drawing to create the isometric box for the curve. Along the front face of the isometric box, transfer dimension A to locate and draw lines 1-2,3-4,5-6, 7-8, 9-10, and 11-12.

Step 3. From points.2, 4, 6, 8, 10, and 12, draw isometric lines on the top face parallel to the D line. Then, measure the horizontal spacing between each of the grid lines in the top multiview as shown for dimension 8, and transfer those distances along isometric lines parallel to the W line. The intersections of the lines will locate points 13 through 18.

Step 4. Draw the curve through points 13 through 18, using an irregular curve. From points 13 through 18, drop vertical isometric lines equal to dimension H. From points 1, 3, 5, 7, 9, and 11, construct isometric lines across the bottom face to intersect with the vertical lines dropped from the top face, to locate points 19 through 24. Connect points 19 through 24 with an irregular curve.

Step 5. Erase or lighten all construction lines to complete the view.

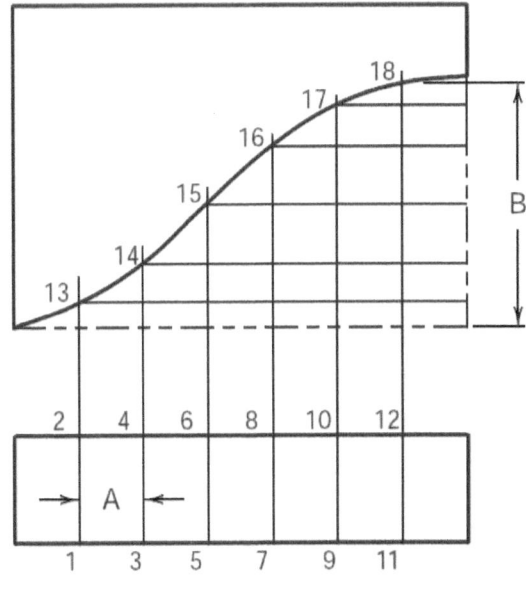

Step 1

77

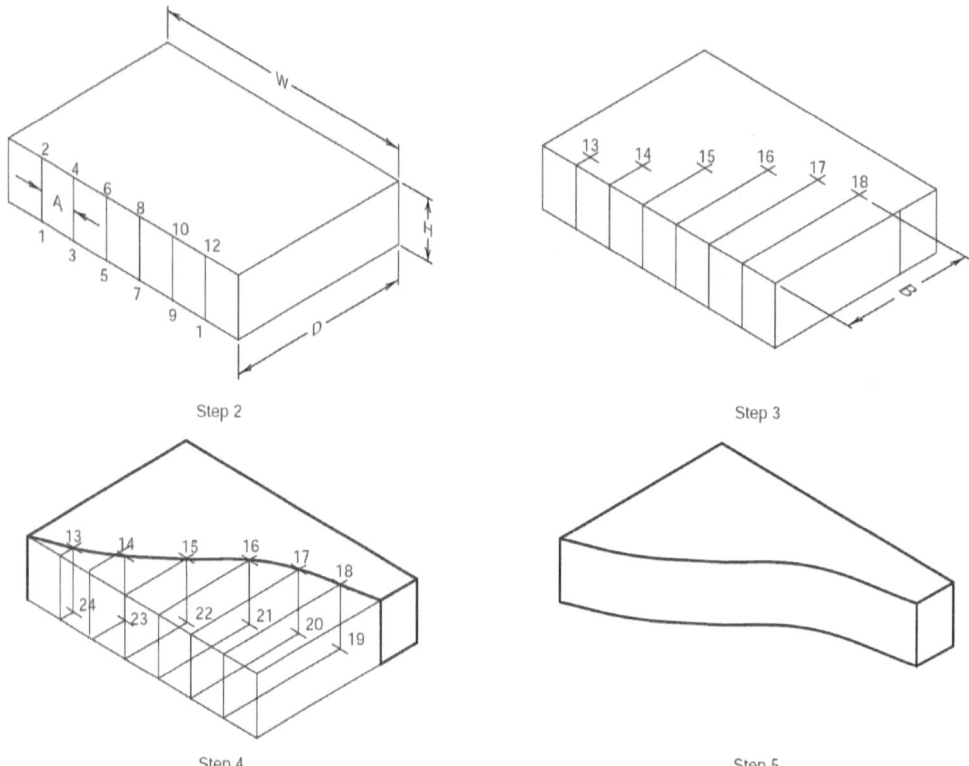

Step 2   Step 3

Step 4   Step 5

Fig. 4.19 Constructing Irregular Curves in an Isometric Drawing

4.3.7 Circular Features

A circular feature will appear elliptical if the line of sight is neither perpendicular nor parallel to the circular face. Circles that lie on any face of an isometric cube will appear as ellipses with 35 degrees 16-minute exposure, as shown in Fig. 4.20. In Fig. 4.21, the center lines for a circle, the axis of a cylinder, and the major and minor axes of an ellipse are drawn for the front, right side, and top planes in an isometric drawing. Note that the minor diameter of the ellipse always coincides with the axis of the cylinder, and the major diameter of the ellipse is always at right angles to the minor diameter. Also, the center lines for the circle never coincide with the major or minor axis of the ellipse.

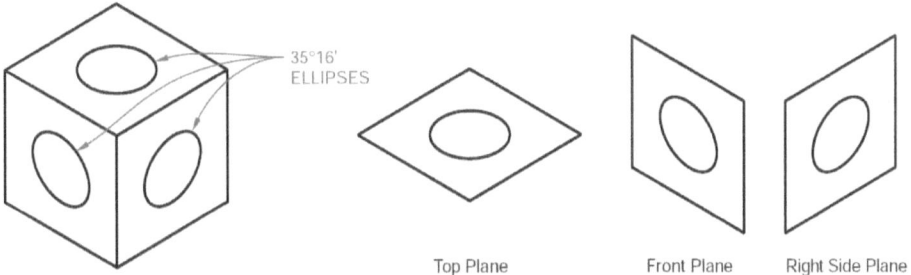

Top Plane   Front Plane   Right Side Plane

Fig. 4.20 Representing Circular Features in an Isometric Drawing

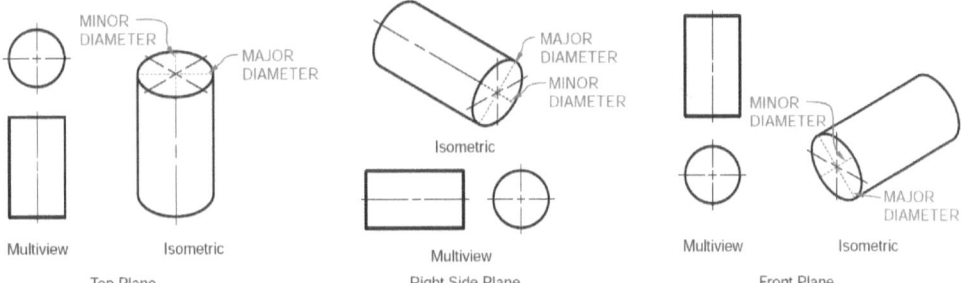

Fig. 4.21 Location of center lines and the major and minor axes of isometric ellipses

Notice that the minor axis is always coincides with the long axis of the cylinder and the major axis is perpendicular to the long axis.

4.3.8 Ellipses

Isometric ellipses representing circular features are drawn using one of three methods:

1. True ellipse construction.
2. Approximate four-center ellipse construction.
3. Isometric ellipse templates.

A true ellipse is constructed by drawing multiple randomly spaced, parallel and perpendicular lines across the circle in the multiview drawing. The lines form a grid similar to the one created for drawing an irregular curve in isometric. Fig. 4.22 and the following steps demonstrate using true ellipse construction to draw a short isometric cylinder.

Step 1. Construct multiple randomly spaced lines parallel to the vertical center line of the circle in the multiview drawing. Construct horizontal lines through the points of intersection on the circumference of the circle. Label all the points of intersection on the circle 1 through 12.

Step 2. Draw the isometric box, using the diameter of the circle as the sides and distance D as the depth dimension. Transfer the measurements from the multiview drawing to the isometric drawing. Construct isometric lines through the points to create a grid and locate points at the intersections. These points lie along the ellipse.

Step 3. Connect the points with an irregular curve to create a true isometric ellipse.

Step 4. Draw isometric lines equal to distance D from each point to create that part of the ellipse which can be seen in the isometric drawing. Use an irregular curve to connect the points, and draw tangent lines to represent the limiting elements of the cylinder.

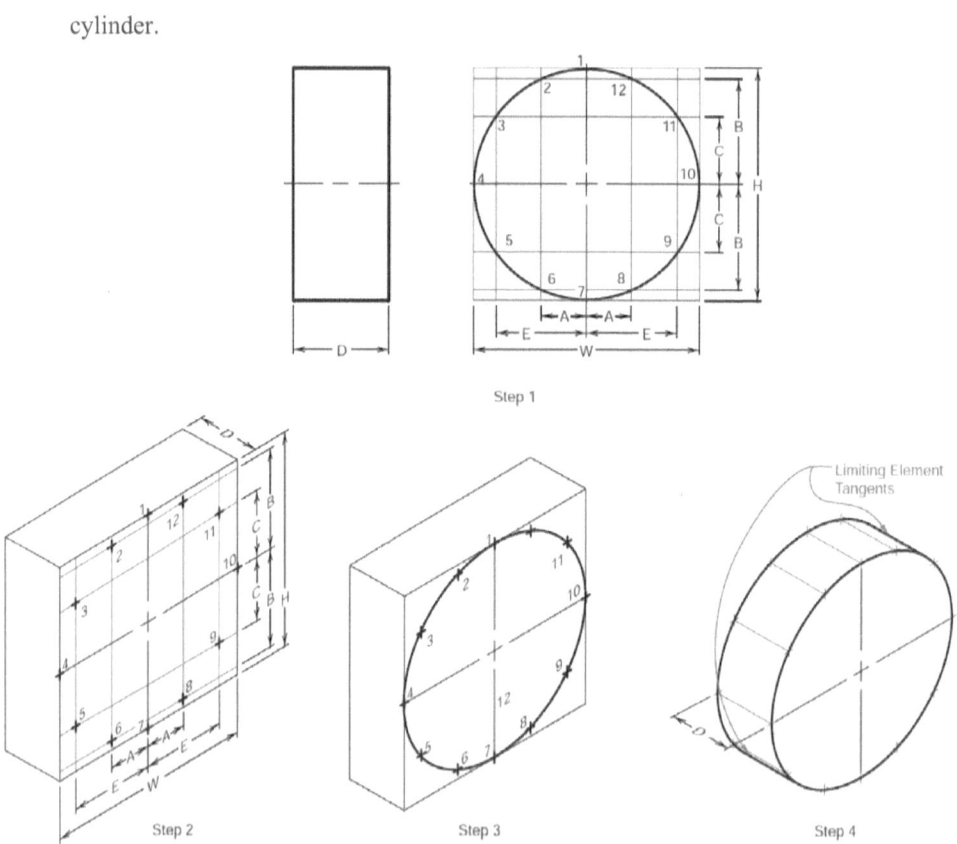

Fig. 4.22 Drawing a true ellipse

80

The four-center construction technique creates an approximate isometric ellipse by creating four center points from which arcs are drawn to complete the ellipse. This method is sufficiently accurate for most isometric drawings. Fig. 4.24 demonstrates the four-center ellipse construction method to construct an isometric cylinder.

Step 1. On the front plane, construct an isometric equilateral parallelogram whose sides are equal to the diameter of the circle.

Step 2. Find the midpoint of each side of the parallelogram. From the midpoint of each side, draw a line to the closest endpoint of the opposite side. These lines are perpendicular to the sides and they intersect at two points, which are the centers for two arcs used to create the approximate ellipse. The perpendicular bisect points are points of tangency for the ellipse.

Step 3. Draw two small arcs with radius rand centers located at points A and B. Draw two large arcs with radius R, using the two corners C and D as the centers.

Step 4. To complete the cylinder, draw the isometric box. Project arc center points A', B', and D' by drawing isometric lines equal to length X.

Step 5. Using radius R with point D' and radius r with points A' and B', construct the isometric ellipse on the rear face. Add tangent limiting element lines to complete the cylinder.

The four-center ellipse is not the same as a true isometric ellipse, but it is close enough for use on most drawings.

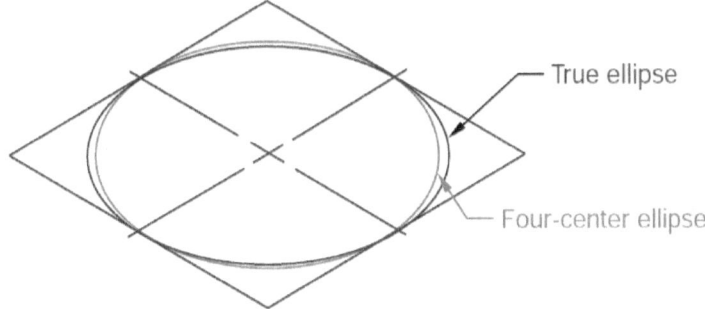

Fig. 4.23 Difference between a True Ellipse and a Four-Center Ellipse

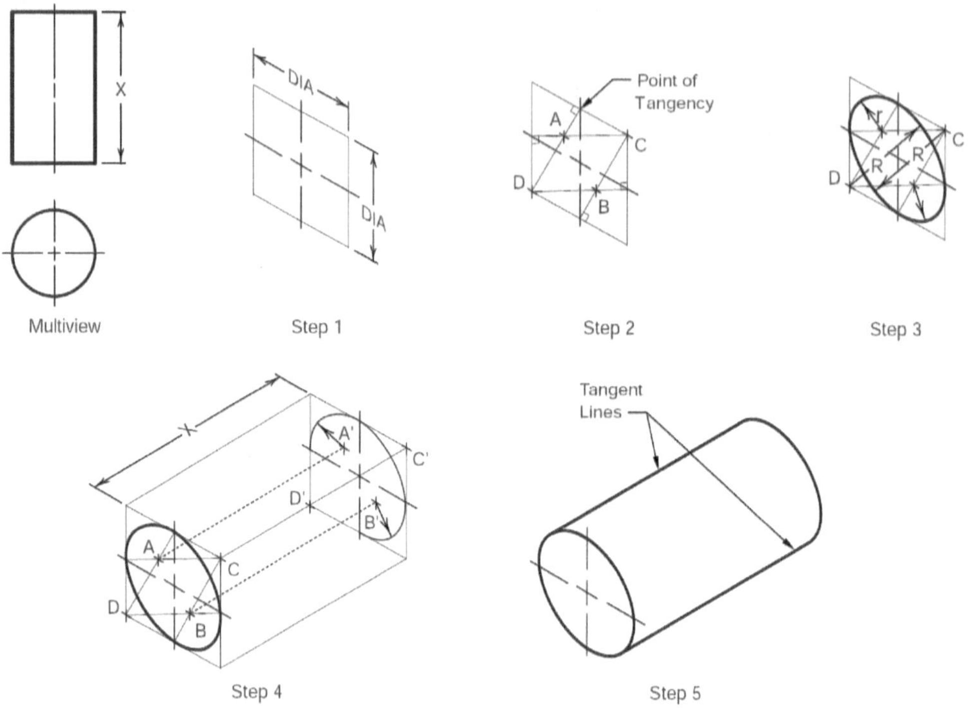

X
DIA
DIA
Point of
Tangency
A
C
D
B
R
R
Multiview
Step 1
Step 2
Step 3
X
A'
C'
D'
B'
A
C
D
B
Tangent
Lines
Step 4
Step 5

Fig. 4.24 Constructing an Ellipse Using the Four-Center Method

Isometric ellipses can also be drawn using **templates.** Make sure the template is an isometric ellipse template, that is, one that has an ***exposure angle of 35 degrees* 16 *minutes.*** Although many different-sized isometric ellipses can be drawn with templates, not every size isometric ellipse can be found on a template. However, it may be possible to approximate a smaller size by leaning the pencil in the ellipse template when drawing the ellipse.

ISOMETRIC ELLIPSE TEMPLATE
XZT-4597-35-16

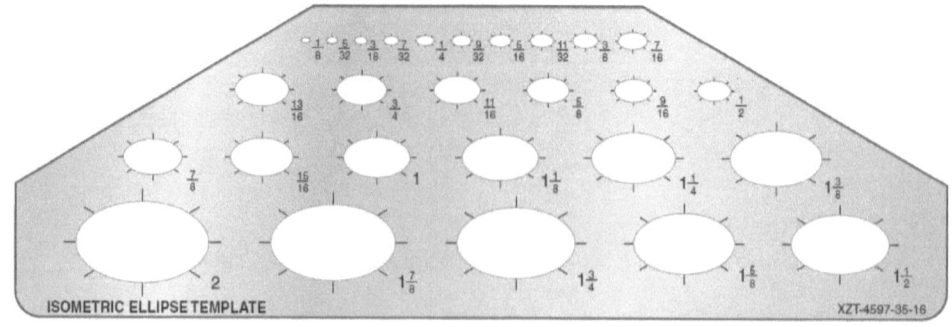

The steps used to draw an isometric ellipse using a template are as follows.

Step 1. The ellipse templates have markings which are used for alignment with the isometric center lines on the drawing (Fig. 4.25). Do not align the ellipse template with the major or minor diameter markings. Locate the center of the circle, then draw the isometric center lines.

Step 2. Find the correct isometric ellipse on the template. Align the isometric distance markings on the ellipse with the center lines, and use a pencil to trace around the elliptical cut in the template.

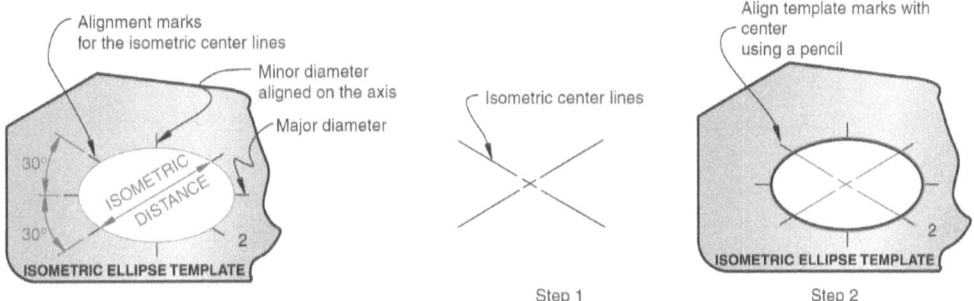

Fig. 4.25 Constructing an ellipse using a template

4.3.9 Inclined Plane Ellipse

An ellipse can be drawn on an inclined plane by constructing a grid of lines on the ellipse in the multiview and plotting points on this grid (Fig. 4.26). The points are then transferred to the isometric drawing, using isometric grid lines. The gridlines produce a series of intersections that are points on the ellipse. These points are then connected with an irregular curve to create an ellipse on the inclined surface. Do not attempt to use an isometric ellipse template, as the exposure angle for this type of ellipse is not 35 degrees 16 minutes.

The following steps describe how to draw an ellipse on an inclined plane.

Step 1: Draw randomly spaced lines parallel to the centerline in the front view of the multiview drawing. Project those lines into the top view. Mark the intersections with the circumference of the circle. In the top view, draw lines parallel to the horizontal center line and through the intersections of the projected lines with the circumference of the circle. You may find it helpful to label the lines, intersection points, and distances between

83

horizontal parallel lines (such as A, S, C) and vertical parallel lines (such as X, Y, R).

Step 2: Construct the isometric box, and then draw the inclined plane in the box. Using distances X, Y, and R, construct isometric lines on the front face of the box. From their intersections with the inclined edge, draw isometric lines across the inclined plane.

Step 3: Transfer distances A, 8, and C to the top edge of the inclined plane and project them across the inclined plane, using nonisometric lines that are parallel to the edges of the inclined plane. The intersections of the two sets of lines on the inclined plane are points on the ellipse.

Step 4: Use an irregular curve to draw the ellipse through the intersection.

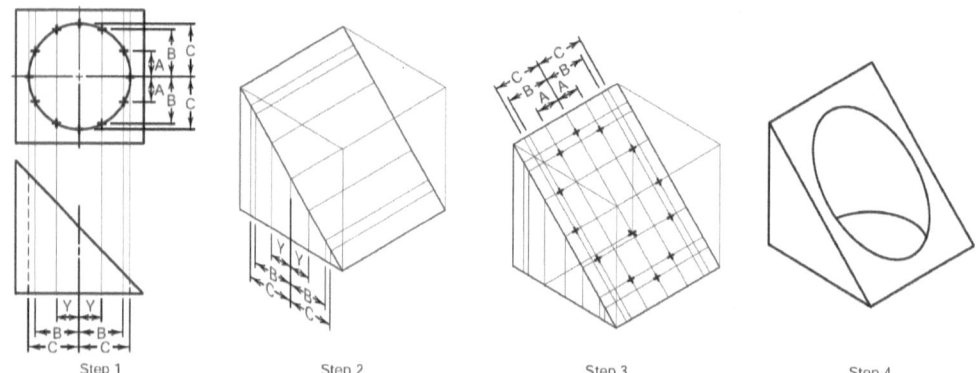

Step 1   Step 2   Step 3   Step 4

Fig. 4.26 Constructing an Ellipse on an inclined Plane

4.3.10 Constructing Arcs in Isometric Drawings

Since arcs are partial circles, they appear in isometric drawings as partial isometric ellipses. Therefore, isometric arcs are constructed using any of the three techniques described for drawing ellipses: true ellipse, four center, or template. The four-center ellipse technique is used in Fig. 4.27.

Step 1. Construct the isometric box, and then draw the isometric parallelogram with the sides equal to the diameter (D) of the arc.

Step 2. Construct the perpendicular bisectors in the isometric parallelogram. Construct the isometric half-circle using radii Rand r from centers 1 and 2.

84

Step 3. Project center points 1 and 2 along isometric lines a distance equal to the thickness (T) of the object. Using the same radii Rand r, construct arcs to create the partial isometric ellipse on the back plane.

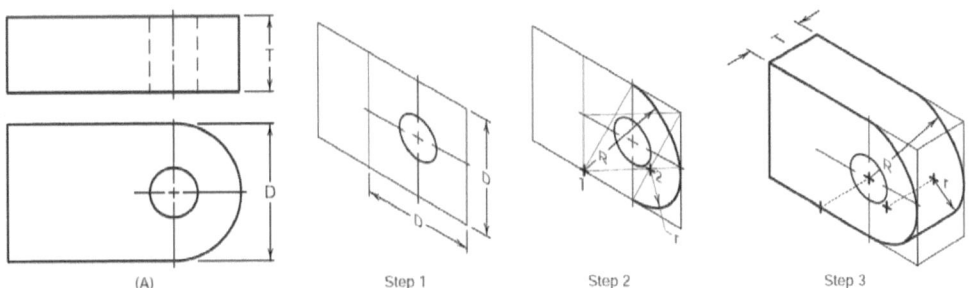

(A) Step 1 Step 2 Step 3

Fig. 4.27 Constructing Arcs in Isometric Drawings

4.3.11 Constructing an Ellipse on an Oblique Surface

Step 1. Using the technique demonstrated in Fig. 4.28, construct a series of grid lines on the oblique surface in the multiview drawing, marking the intersection points representing the hole.

Step 2. Transfer the grid lines to the oblique surface in the isometric drawing.

Step 3. Mark the intersection points of the grid lines and use an irregular curve to connect these points, creating the curved intersection between the hole and the oblique plane.

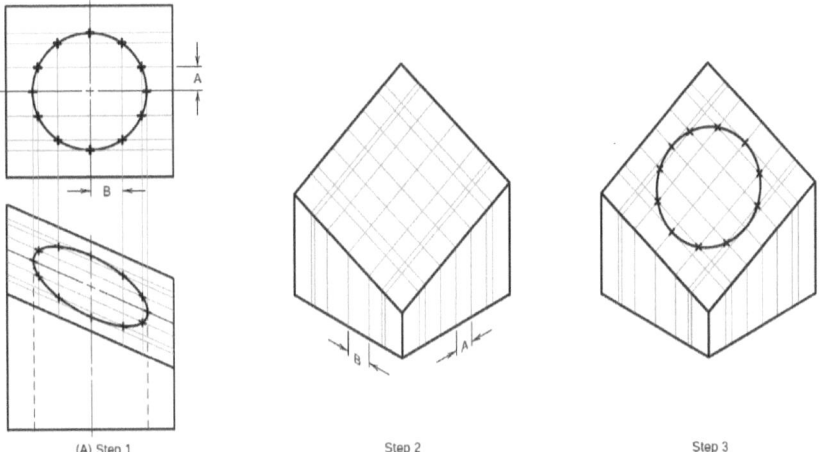

(A) Step 1 Step 2 Step 3

Fig. 4.28 Constructing an Ellipse on an Oblique Surface

85

4.3.12 How to center an Isometric Object

An isometric drawing should be located in the center of a drawing space. Follow the following steps to locate the drawing in the center of the drawing space.

Step 1. Locate the center point of the work area and do the following: go North' 1/2 H, go s60°W a distance of 1/2 D, and go S60oE, 1/2 W to establish a starting point,

Step 2. From the starting point, draw the basic shape of the object

Step 3: construct the object within the basic shape

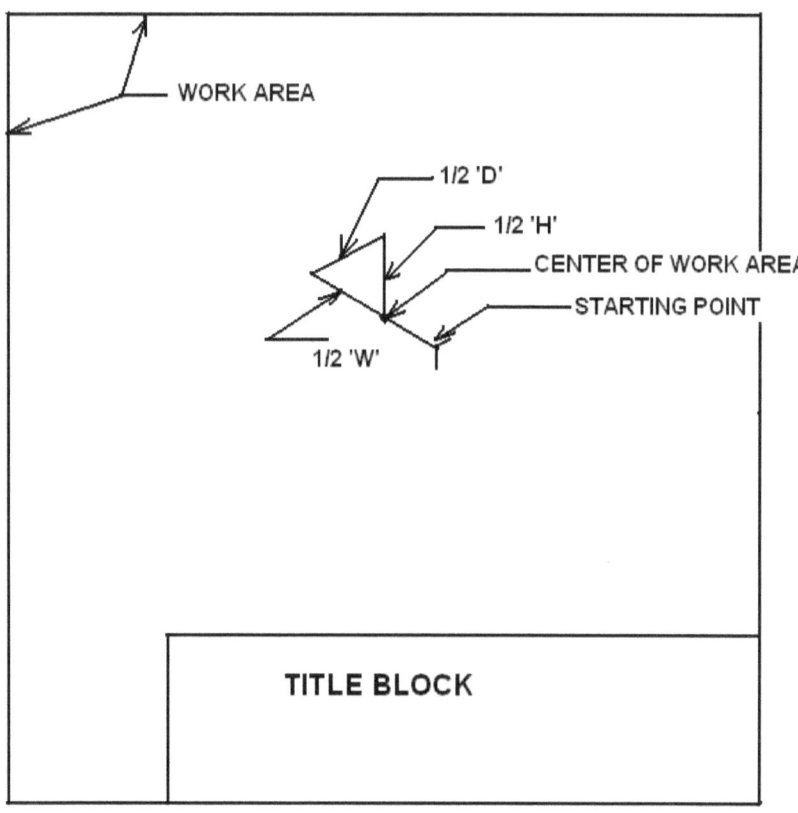

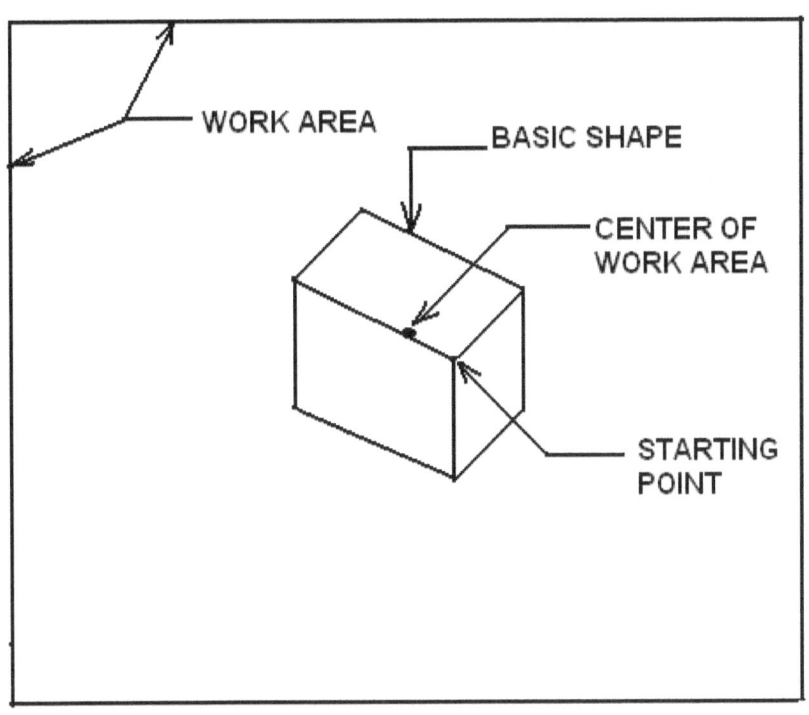

WORK AREA

BASIC SHAPE

CENTER OF
WORK AREA

STARTING
POINT

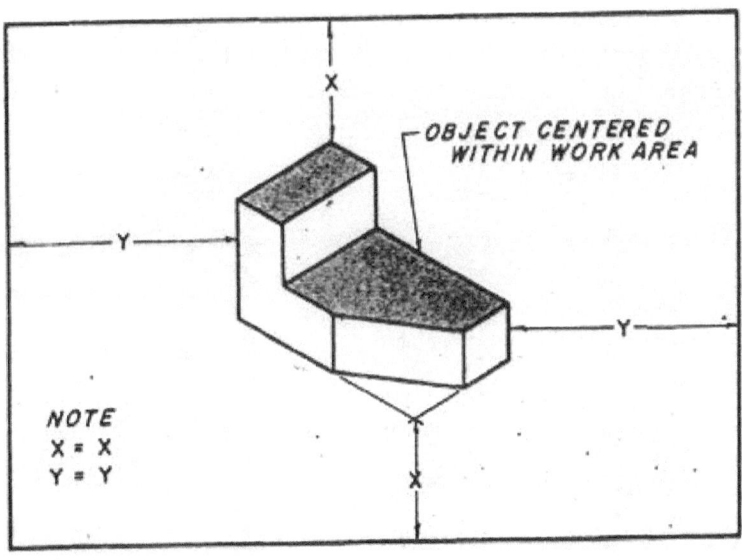

X

OBJECT CENTERED
WITHIN WORK AREA

Y

Y

NOTE
X = X
Y = Y

X

Fig. 4.29 Locating an isometric drawing at the center of drawing space

4.4 Oblique Drawing

Oblique drawings are a form of pictorial drawing in which the <u>most descriptive or natural</u> <u>view is</u> <u>treated as the front view and is placed parallel to the plane of projection</u>. For example, oblique drawing is the pictorial method favored by the furniture manufacturing and cabinet making industry. However, because of the excessive distortion that occurs, oblique drawings are not used as commonly as other types of pictorials.

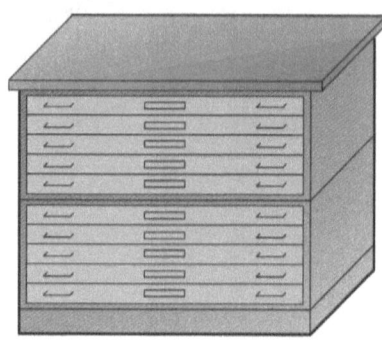

Fig. 4.30 Typical furniture industry oblique drawing

4.4.1 Oblique Projection Theory

Oblique projection is the basis for oblique drawing and sketching. Oblique projection is a form of parallel projection in which the projectors are parallel to each other but are not perpendicular to the projection plane.

The actual angle that the projectors make with the plane of projection is not fixed; thus, different angles can be used. However, angles of between 30 degrees and 60 degrees are preferable because they result in minimum distortion of the object.

If the principal view of the object is placed parallel to the projection plane and the line of sight is something other than perpendicular to the projection plane, the resulting projection is an oblique pictorial. The path of the projectors follows the receding edges of the object.

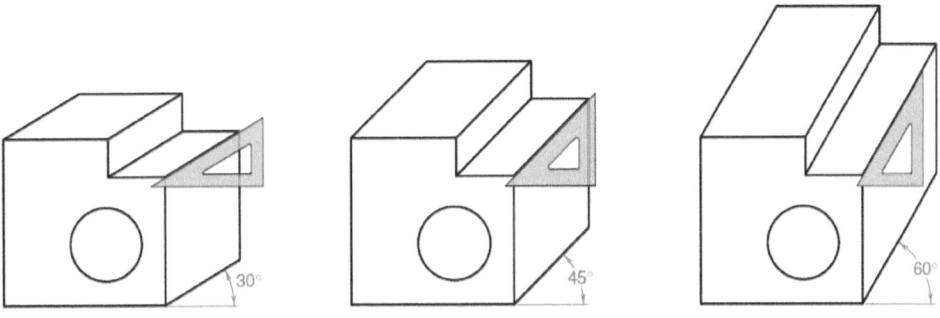

Fig. 3.31 Oblique drawing angles. Typical oblique drawing angles used are 30, 45, or 60 degrees from the horizontal

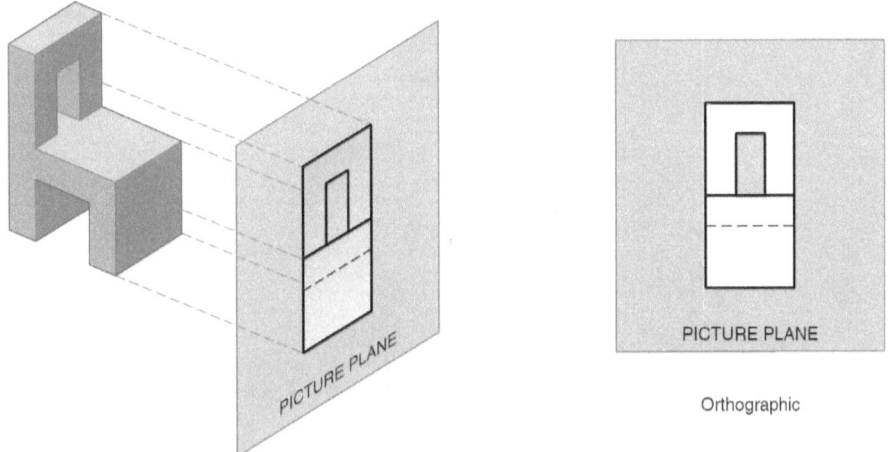

Fig. 3.32 Orthographic projection. In orthographic projection, the projectors are perpendicular to the projection plane.

4.4.2 Oblique Drawing Classifications

There are three basic types of oblique drawings: (1) *cavalier*, (2) *cabinet*, and (3) *general*. All three types are similar in that their front surfaces are drawn true size and shape and parallel to the frontal plane. The receding angles can be anywhere from 0 to 90 degrees, although angles of less than 45 degrees or greater than 60 degrees cause extreme distortion. The difference between the three types relates to the measurements made along the receding axis. The cavalier oblique is drawn true length along the receding axis. The cabinet oblique is drawn half the true length along the receding axis.

The general oblique can be drawn anywhere from full to half-length along the receding axis.

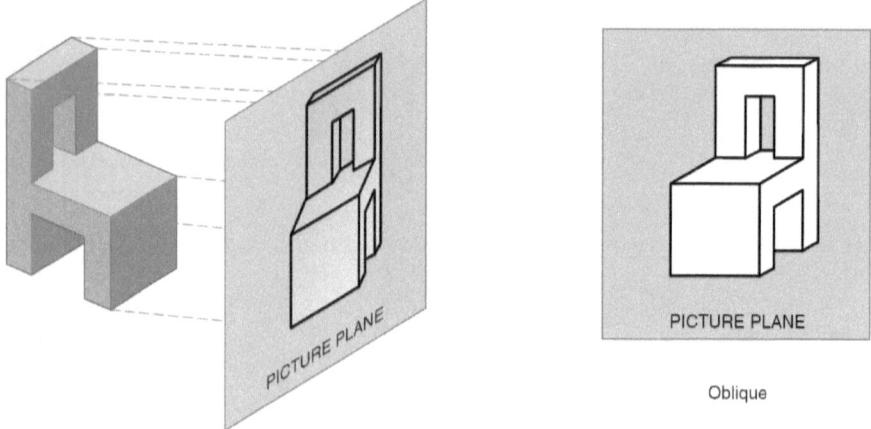

Oblique

Fig. 4.33 Oblique projection. In oblique projection, the projectors are never perpendicular to the projection plane.

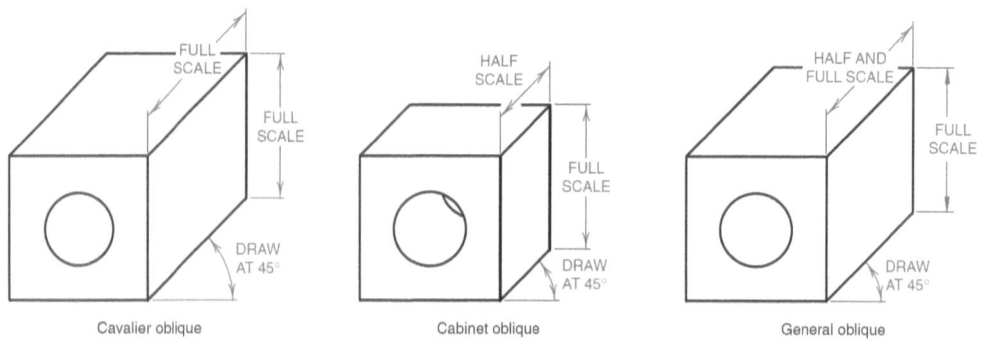

Cavalier oblique   Cabinet oblique   General oblique

Fig. 4.34 Types of oblique drawings

The half-size receding axis on the cabinet oblique reduces the amount of distortion; therefore, the cabinet oblique drawing is the one used for most illustrations. The cabinet oblique foreshortens the receding axis and gives a more realistic view.

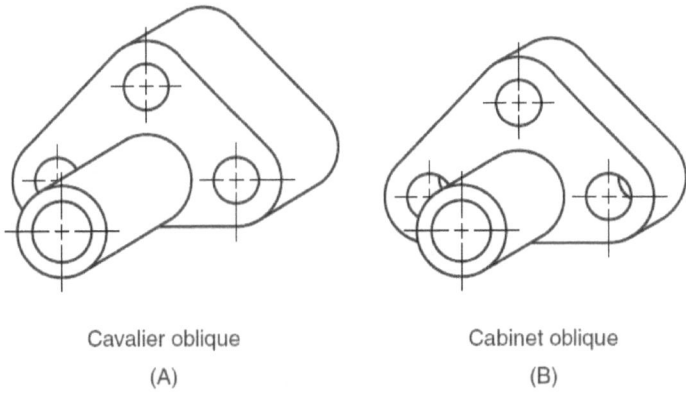

Cavalier oblique

Cabinet oblique

(A)

(B)

Fig. 4.35 Comparison of cavalier and cabinet oblique drawings

4.4.3 Object Orientation Rules

In oblique projection, the object face that is placed parallel to the frontal plane will be drawn true size and shape. Thus, the first rule in creating an oblique drawing is to <u>place complex features, such as arcs, holes, or irregular surfaces, parallel to the frontal plane</u>. This allows these features to be drawn more easily and without distortion. Place holes and arcs parallel to the frontal plane whenever possible to avoid distortion and to minimize having to draw circles as ellipses.

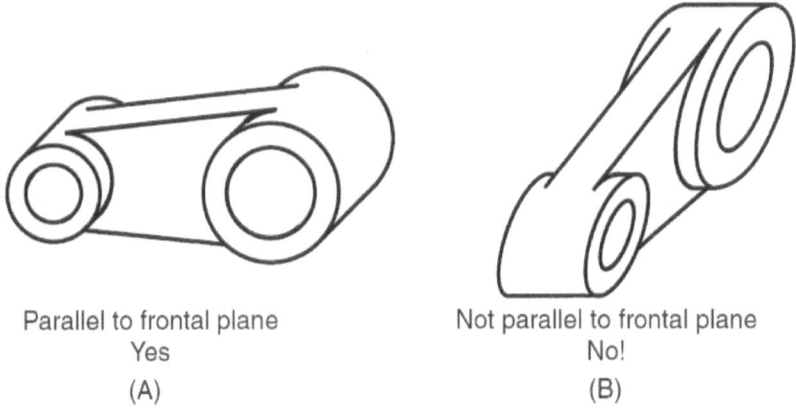

Parallel to frontal plane
Yes

(A)

Not parallel to frontal plane
No!

(B)

Fig. 4.36 Object orientation

The second rule in developing oblique drawings is that the longest dimension of an object should be parallel to the frontal plane. If there is a conflict between these two rules, the first rule takes precedence because representing complex geometry without distortion is more advantageous.

91

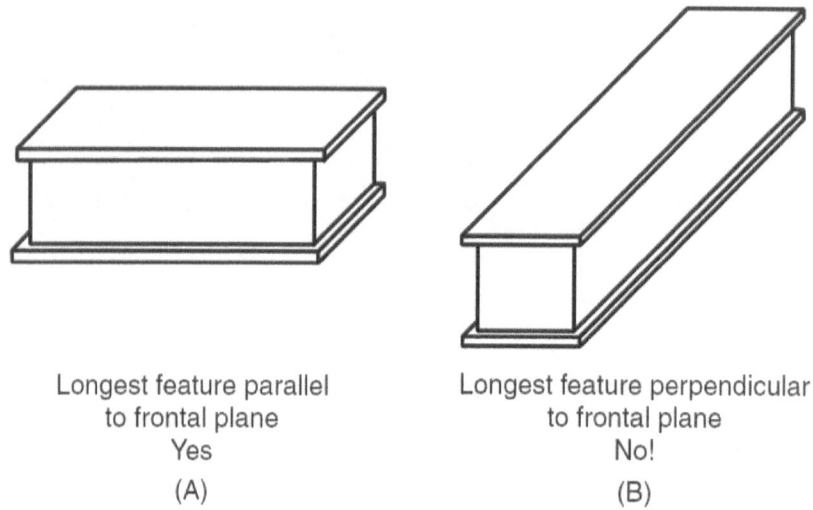

Longest feature parallel
to frontal plane
Yes
(A)

Longest feature perpendicular
to frontal plane
No!
(B)

Fig. 4.37 Long dimension orientation. Place the longest dimension of the object parallel to the frontal plane to avoid distortion.

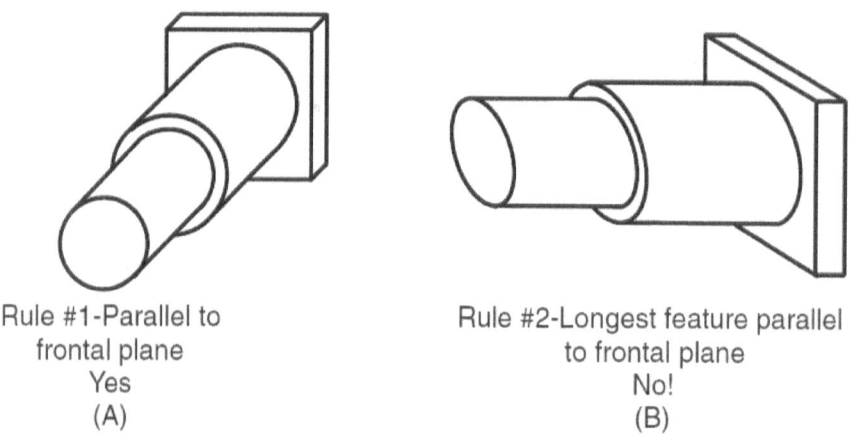

Rule #1-Parallel to
frontal plane
Yes
(A)

Rule #2-Longest feature parallel
to frontal plane
No!
(B)

Fig. 4.38 Cylinder rule. The cylinder rule overrides the longest-dimension rule when creating an oblique drawing.

4.4.4 Oblique Drawing Construction

To develop an oblique drawing that consists mostly of prisms, use the box construction technique. Determine from the given multiviews the best view for the front face of the oblique drawing. Decide what type of oblique to draw, and define the angle for the receding lines. For this example, the front view of the object will be the front face, the cavalier oblique will be used, and the receding angle will be 45 degrees.

If the object is composed mostly of full or partial cylindrical shapes, place these shapes in the frontal plane so that they will be drawn true size and shape.

Step 1. Determine the overall height, width, and depth dimensions, and begin to create the construction box. Draw the receding axes at 45 degrees, and layout the depth dimensions (in true length).

Step 2. Draw the front view, which is identical to the orthographic front view, using dimensions M, N, 0, and P. Note that the front surface with the two holes is drawn in the frontal plane of the oblique construction box. The other parts of the front view must be drawn at a distance P behind the frontal plane of the construction box. Distance P is measured along one of the 45-degree receding lines. Other details are added to the drawing to complete the construction.

Step 3. Darken or trim the lines to complete the drawing.

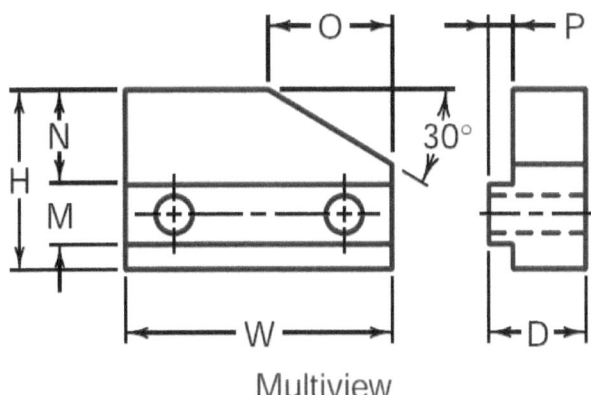

Multiview

93

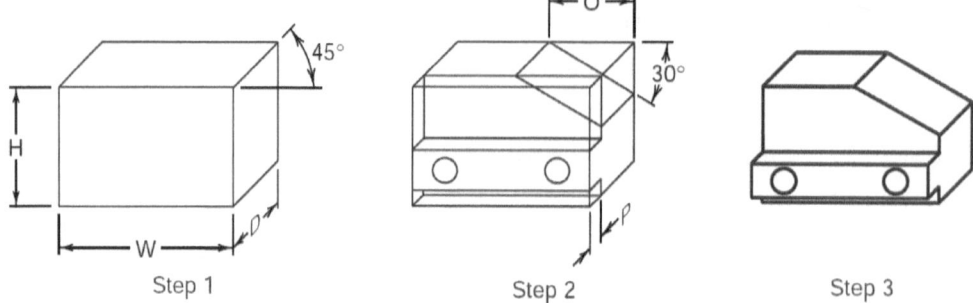

Fig. 4.39 Constructing an Oblique Drawing Using the Boxing-In Method

4.4.5 Constructing an Oblique Drawing of an Object with Circular Features

From the given multiviews, determine the best view for the front face of the oblique drawing. Decide what type of oblique to draw, and define the angle for the receding lines. For this example, the front view of the object will be the front face, the cavalier oblique will be used, and the receding angle will be 45 degrees. Again, depth dimensions are drawn full- scale on the receding lines.

Step **1.** Block in the overall width, height, and depth dimensions, and use 45 degrees for the receding lines.

Step **2.** On the front face, locate the center A of the 1" radius arc and 0.625 diameter hole. Draw the arc and the hole on the front face of the oblique drawing.

Step **3**. Locate center B of the 1" arc and the hole in the back plane of the oblique drawing by projecting point A along a line at a 45-degree angle and 1" long (the depth dimension). Draw another 1" radius arc using point B as the center. Determine points of tangency by constructing vertical lines from the centers to the top edges and other lines perpendicular to the receding line that is the bottom edge of the right side of the construction box.

Step **4**. Mark the notch dimensions on the front face, and mark the depth of the notch along the top left receding line. Draw the notch using the dimensions marked on the edges of the box.

Step **5**. Locate the center of the 0.25" radius arc, and mark the location of the slot. Draw the arc and then the slot, and project point C back at a 45-degree angle and *OS'* long. Draw the back part of the slot that is visible

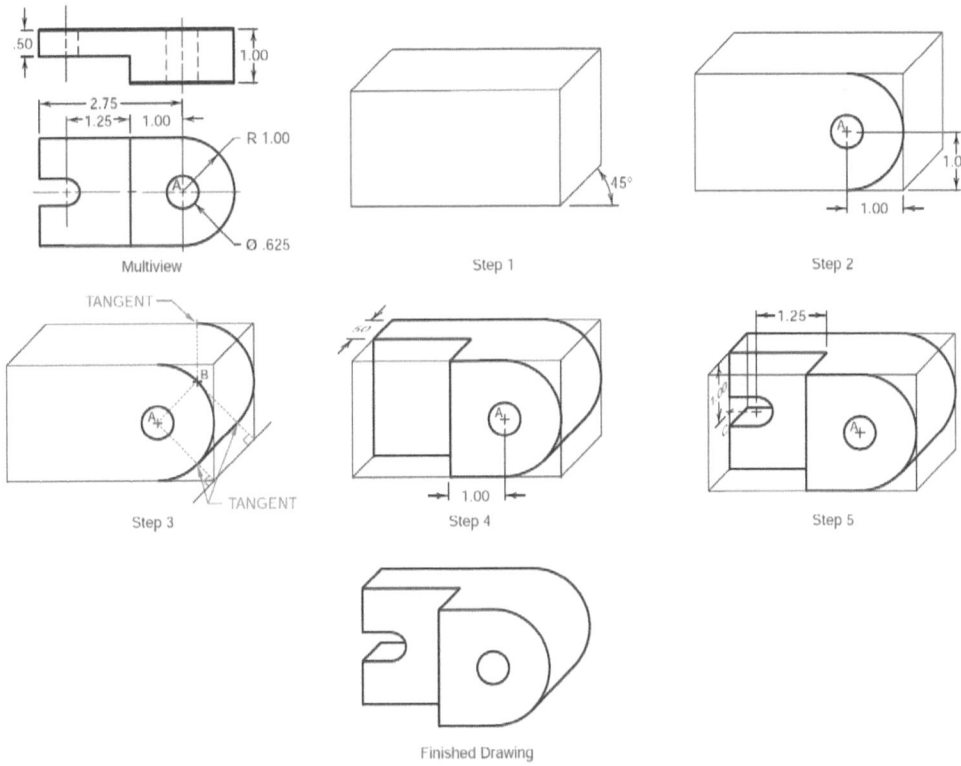

Multiview   Step 1   Step 2

Step 3   Step 4   Step 5

Finished Drawing

Fig. 4.40 Constructing an Oblique Drawing of an Object with Circular Features

4.4.6 Construction of Circles in Oblique Drawing

It is not always feasible to position an object such that all of its cylindrical features are parallel to the frontal plane. In oblique drawing, an *alternate four-center ellipse method* can be used to draw circles not in the frontal plane. However, this method can only be used in cavalier oblique drawings because the receding axes are drawn full scale; thus, the parallelogram used to develop the ellipse is an equilateral parallelogram. The regular four- center method used in isometric drawing will only work with oblique drawings that have receding axes at 30 degrees (the same axis angle used in isometric drawings). Any other receding axis angles call for using the alternate four-center ellipse method described in the following steps.

Step 1. On the surface that is not parallel to the frontal plane, locate the center of the circle, and draw the center lines.

Step 2. From the center point, draw a construction circle that is equal to the diameter of the circle that is being constructed. This construction circle will intersect the center lines at four points, A, B, C, and D.

Step 3. From points A and C, construct lines that are perpendicular to center line BD.

Step 4. From points B and 0, construct lines that are perpendicular to center line AC.

Step 5. From the intersections of the perpendicular construction lines, draw four circular arcs using the radii marked.

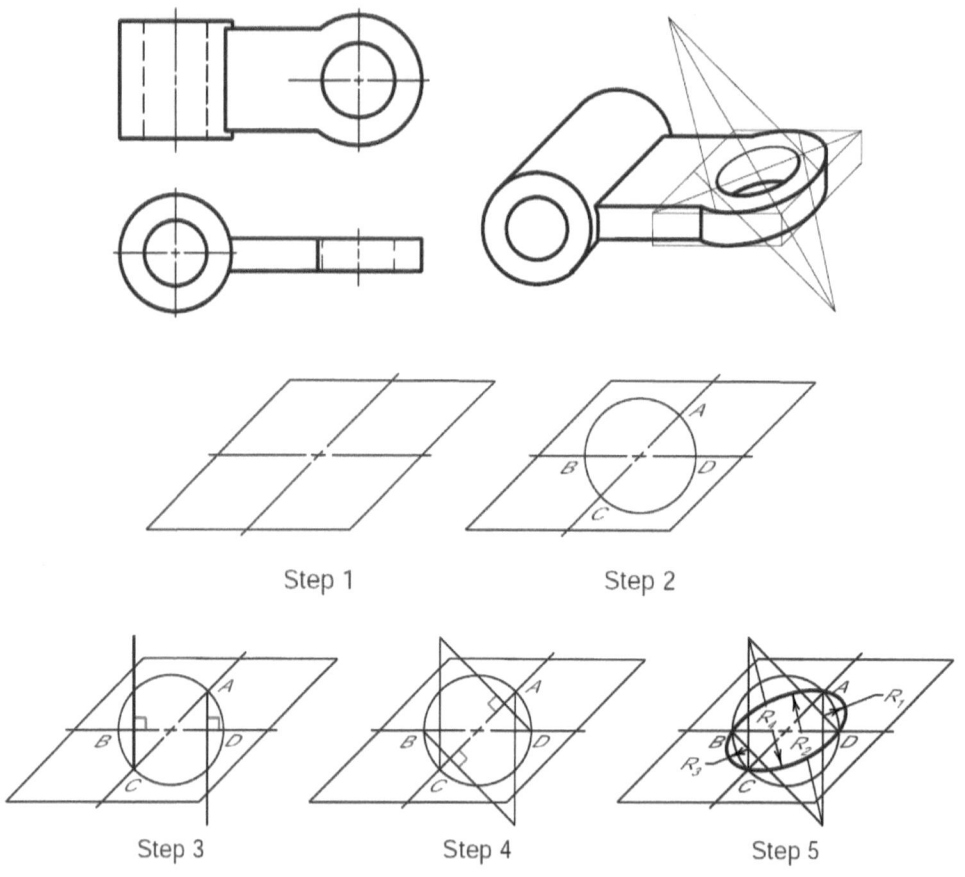

Fig. 4.41 Four-Center Ellipse Construction

4.4.7 Constructing Oblique Curved Surfaces

Cylinders, circles, arcs, and other curved or irregular features can be drawn point-by-point using the **offset coordinate method.** The offset coordinate method of drawing curves is a technique that uses a grid of equally spaced lines on the multiview to divide the curve into parts so that points along the curve can be located and transferred to the oblique view for construction of the oblique curve.

The offset coordinate grid is first drawn on the orthographic view of the object, and the coordinate points delineating the curve are marked as shown in Figure A. The grid and coordinate points are then transferred to the oblique drawing, and the curve is drawn with an irregular curve. If the oblique drawing is a cabinet or general oblique, the offset measurements from the multiview drawing must be transferred at the required reduced scale on the receding axis (Fig. 4.42 B and C).

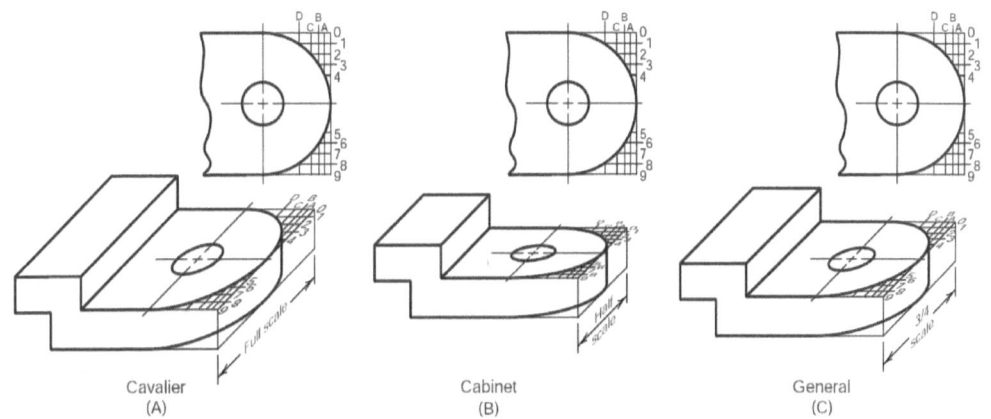

Cavalier
(A)

Cabinet
(B)

General
(C)

Fig. 4.42 Constructing Oblique Curved Surfaces Using Offset Coordinates

The alternate four-center ellipse method cannot be used with oblique drawings that are not full scale; therefore, circular features in these oblique drawings must be constructed using the **offset coordinate method**. This method is also used to construct an ellipse in a plane that is inclined to the frontal plane of the oblique drawing. Horizontal cutting planes are used to locate coordinate points, as shown in Fig. 4.43 B. The cutting planes and coordinate points are then transferred to the oblique drawing (Fig. 4.43 C), and an irregular curve is used to draw the curved surface.

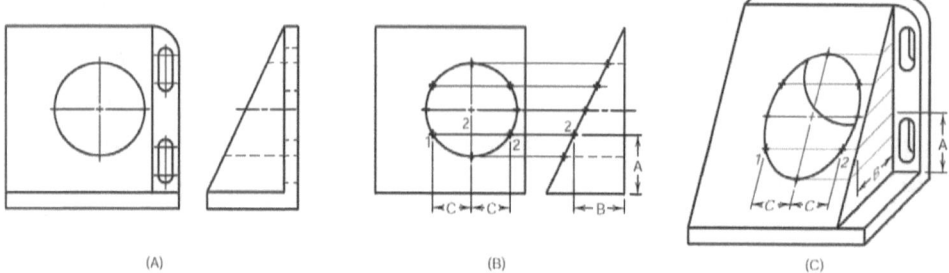

(A)   (B)   (C)

Fig. 4.43 Constructing an Ellipse Inclined to the Frontal Plane Using the Offset Coordinate
Method

4.4.8 Constructing Oblique Angles

True angular measurements can only be made in an oblique drawing when the plane that contains
the angle is parallel to the frontal plane. If the angle lies in one of the oblique receding planes, the
angle must be developed by linear measurements (Fig. 4.44 A). The measurements are then
transferred to the oblique drawing. If the drawing is a cabinet oblique, then all the receding
dimensions must first be reduced to half size before being transferred to the oblique drawing, as
shown in fig. 444 B.

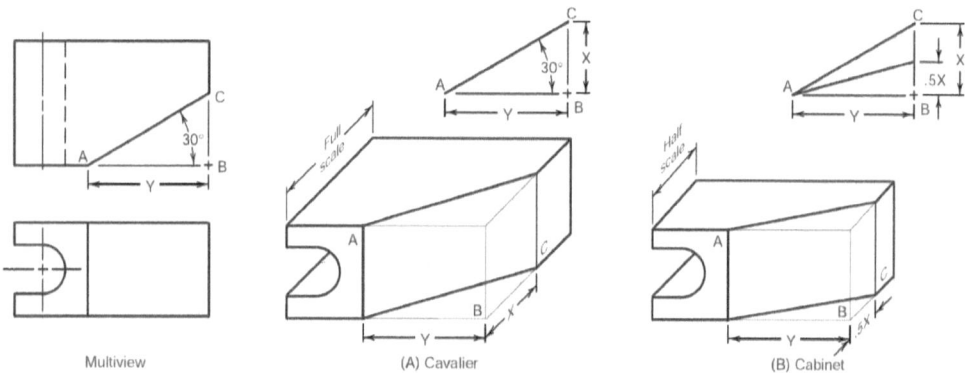

Multiview   (A) Cavalier   (B) Cabinet

Fig. 4.44 Constructing Oblique Angles

4.5 Perspective Drawings

Perspective drawings are a type of pictorial drawing used to represent 3-D forms on 2-D media.
Such drawings create the <u>most realistic representations of objects</u> because the <u>human visual system</u>

98

creates images that closely resemble perspective drawings. To prove this, stand in the middle of a long, flat road. Look at the edges of the road and follow those lines into the horizon. You will see that the lines appear to converge at a common point on the horizon.

One of the most important features of perspective drawings is the **convergence** of parallel edges as they recede from the viewer.

Fig. 4.45 Convergence as seen in a photograph. This photograph shows parallel railroad lines receding to a point on the horizon.

4.5.1   Perspective Projections Terminology

- ✓ The **horizon line** is the position that represents the eye level of the observer.
- ✓ The **station point** in the perspective drawing is the eye of the observer.
- ✓ The **picture plane** is the plane upon which the object is projected.

In Figure below, telephone pole AB is projected onto the picture plane and appears foreshortened as A'B'. Likewise, telephone pole CD is projected onto the picture plane and appears foreshortened as CD'. Object distance relative to the picture plane can be summarized as follows:

1. As objects move further behind the picture plane, they are projected as smaller images.

2. As objects move further in front of the picture plane, they are projected as larger images.

3. Objects positioned *in* the picture plane are shown true size.

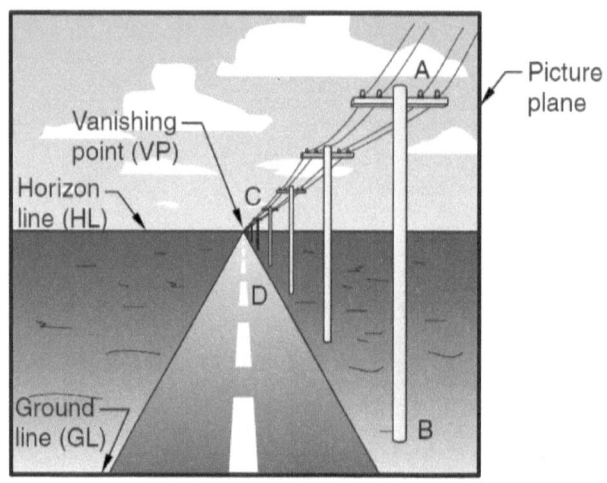

Perspective View

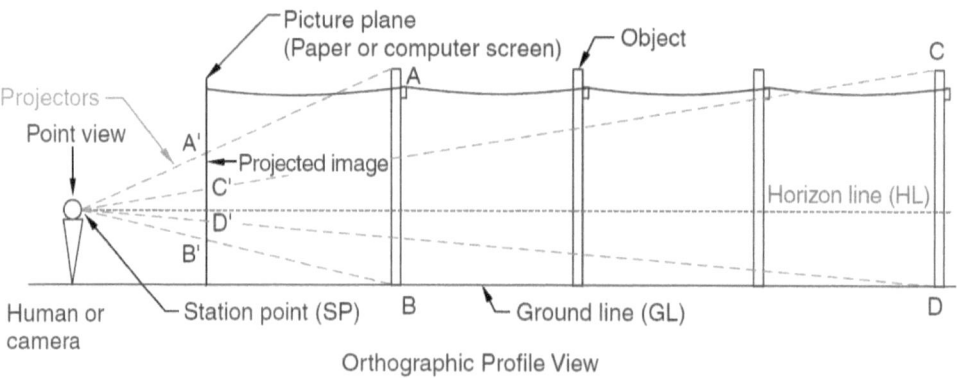

Orthographic Profile View

Fig. 4.46 Perspective and orthographic profile views of a scene

All parallel lines that are *not* parallel to the picture plane, such as the edges of the road in, converge at the vanishing point. All parallel lines that *are* parallel to the picture plane, such as the telephone poles in, remain parallel and do not recede to a vanishing point.

An object positioned at an infinite distance from the picture plane appears as a point, called the **vanishing point**. A vanishing point is the position on the horizon where lines of projection converge. Placing the vanishing point directly behind the object, as shown in above and Fig. 4.47 B, creates a view looking straight at the object. Placing the vanishing point to the right of the object, as shown in Fig. 4.47 A, produces a view showing the right side. Similarly, placing the vanishing point to the left of the object produces a view showing the left side, as shown in Fig. 4.47 C.

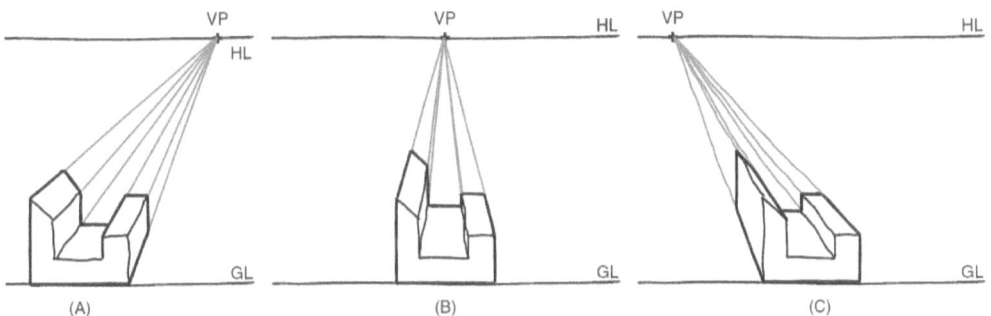

Fig. 4.47 Vanishing point position. Changing the vanishing point changes the perspective view.

The figure below demonstrates how the relationship between the horizon line (HL) and ground line (GL) determines the type of perspective view:

- ✓ **Bird's eye view**: ground line below the horizon line.
- ✓ **Human's eye view**: ground line six feet below the horizon line.
- ✓ **Ground's eye view**: ground line at the same level as the horizon line.
- ✓ **Worm's eye view**: ground line above the horizon line.

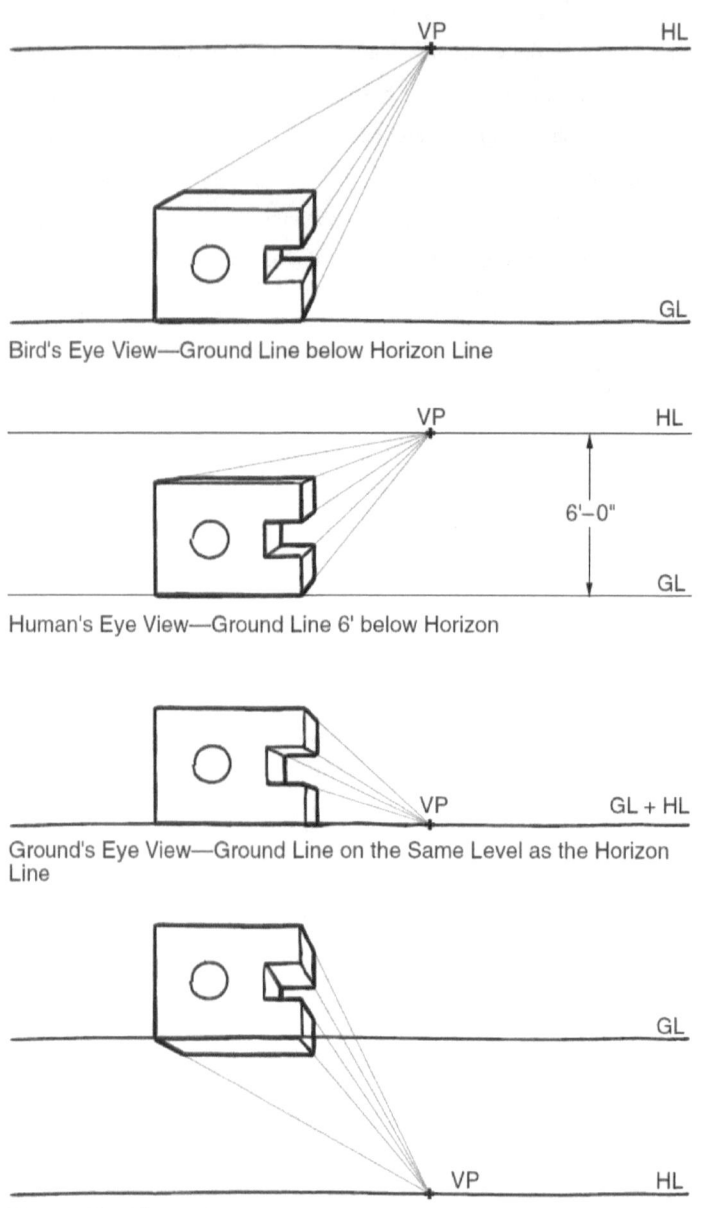

Bird's Eye View—Ground Line below Horizon Line

Human's Eye View—Ground Line 6' below Horizon

Ground's Eye View—Ground Line on the Same Level as the Horizon Line

Worm's Eye View—Ground Line above the Horizon Line

Fig. 4.48 Ground line position. Changing the ground line relative to the horizon line changes the perspective view created

4.5.2 Perspective Projection Classifications

Perspective views are classified according to the number of vanishing points: 1, 2 or 3. Increasing the number of vanishing points increases the realism of the drawing but also increases the drawing difficulty. The vanishing points for one- and two-point perspective drawings both go to the horizon line. The third vanishing point in a three-point perspective drawing is located perpendicular to the horizon line.

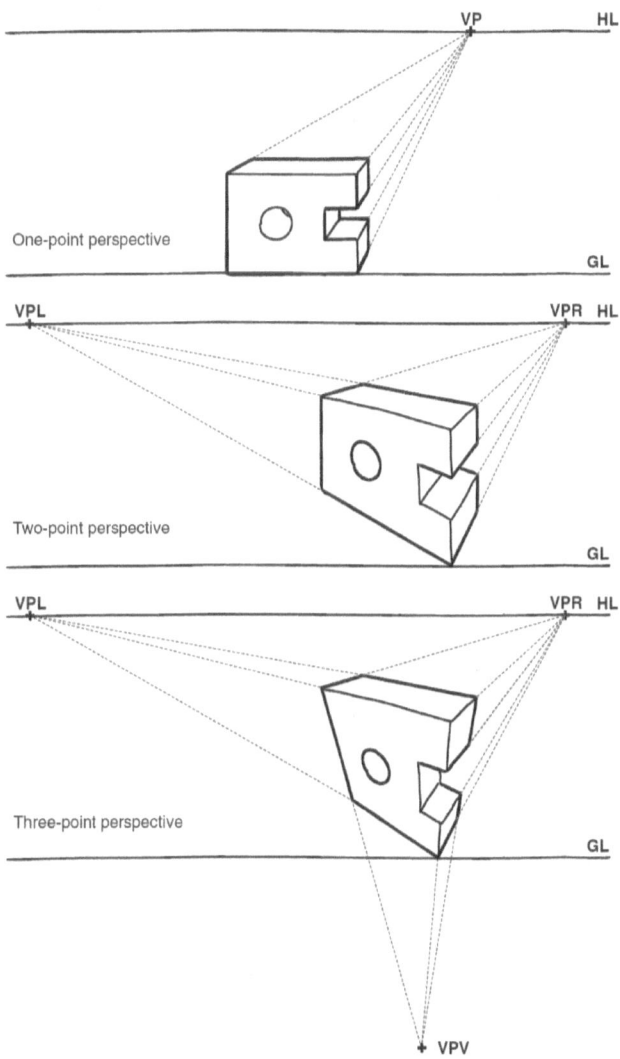

Fig. 4.49 Classification of perspective drawings

4.5.3 One-point Perspective Sketching

1. Sketch front surface of object and locate vanishing point.

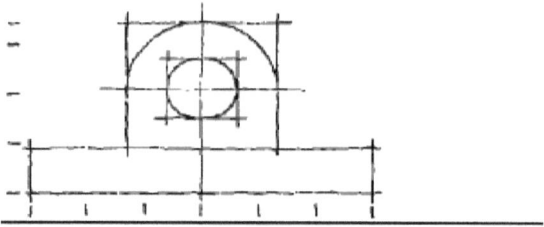

2. Sketch receding lines from intersections and points of tangency on front surface to vanishing point

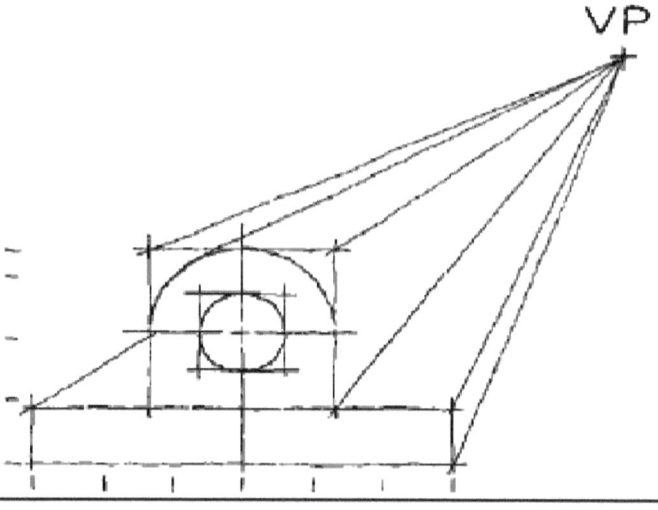

3. Estimate the depth of the object you will show and block in the back surface between the receding lines

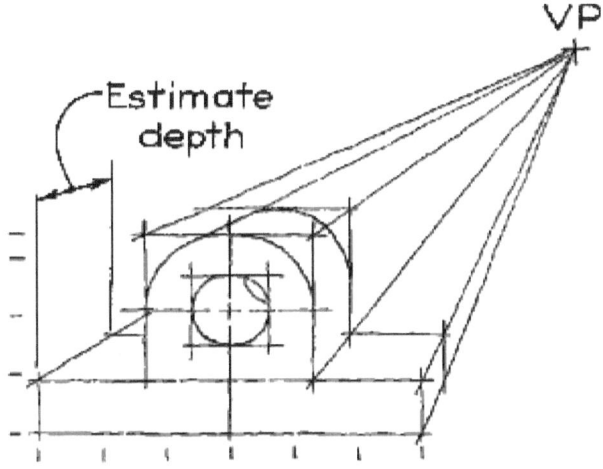

4.6 Problems

a) For the orthographic views given below, create isometric or oblique drawings.

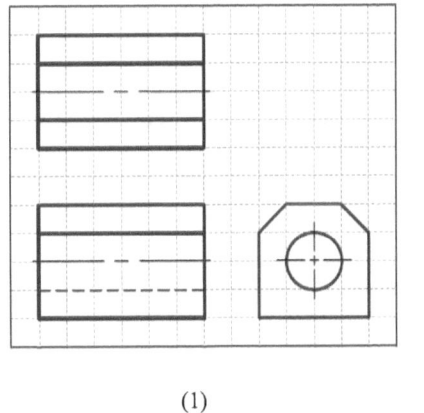

(1)

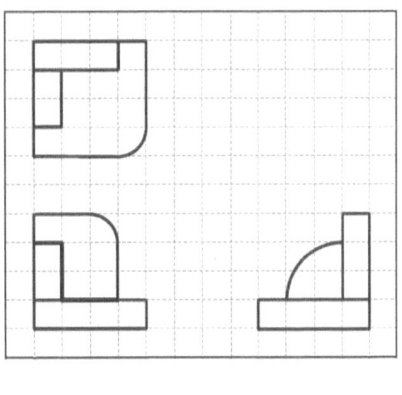

(2)

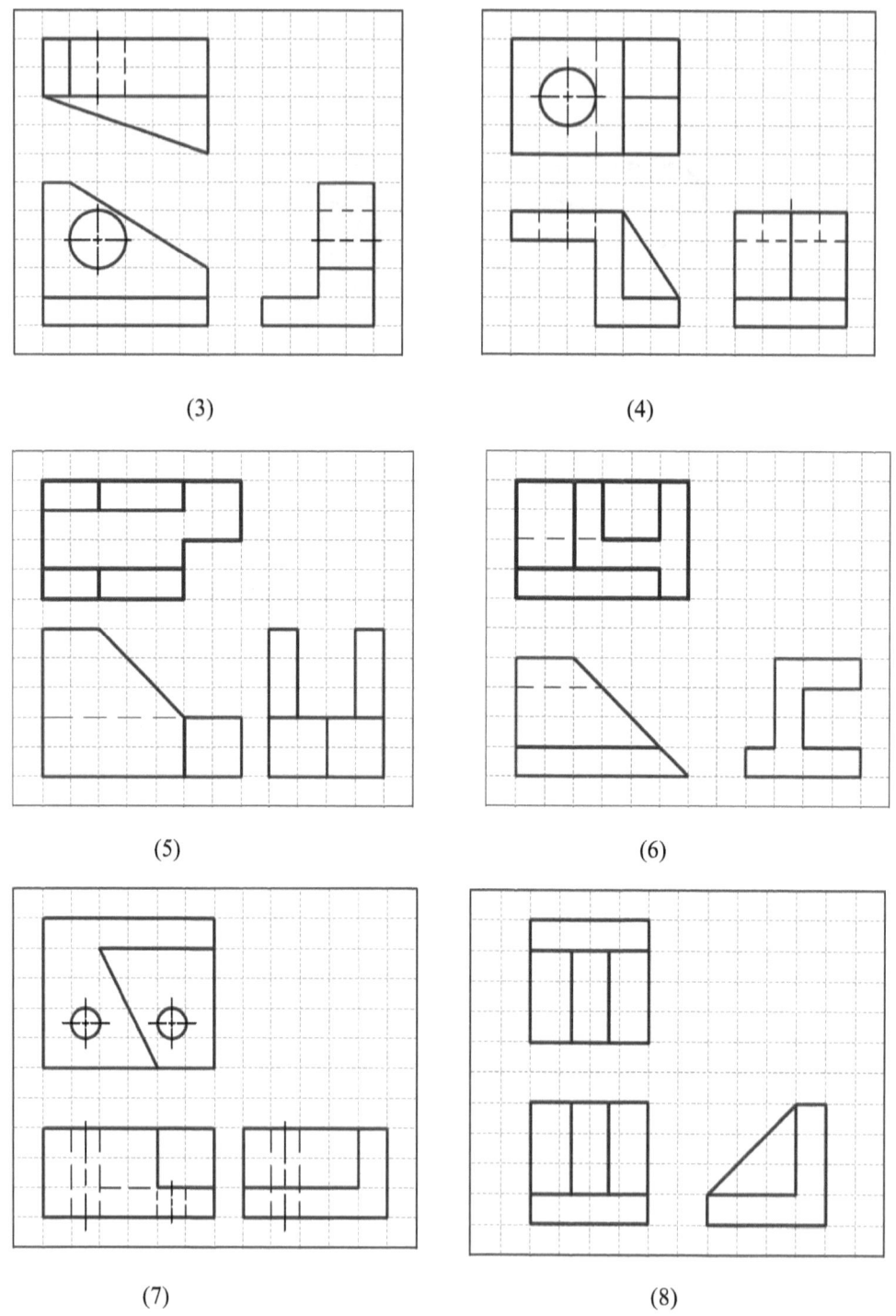

(3)

(4)

(5)

(6)

(7)

(8)

106

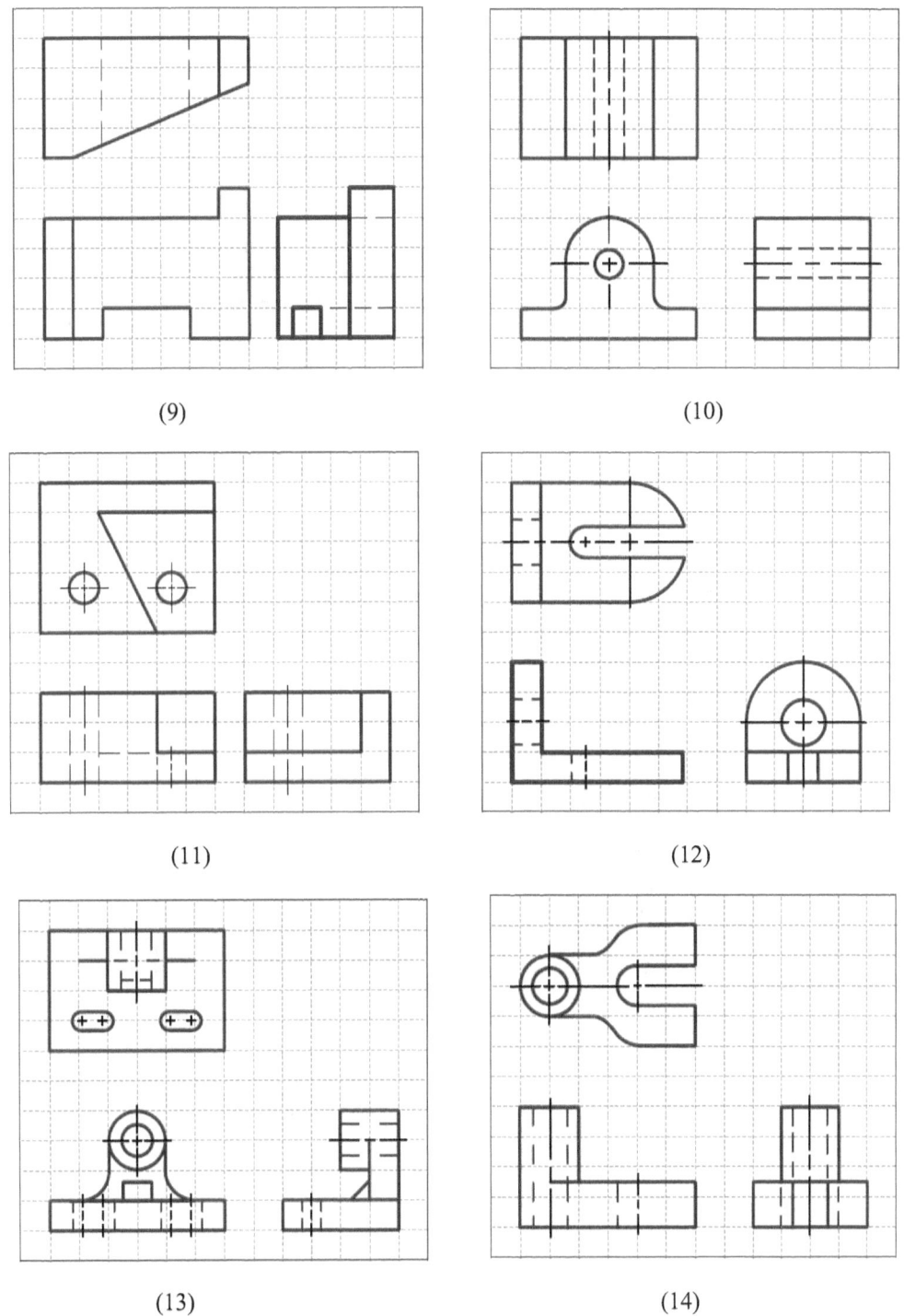

(9)

(10)

(11)

(12)

(13)

(14)

b) Draw isometric Drawing from the given orthographic views.

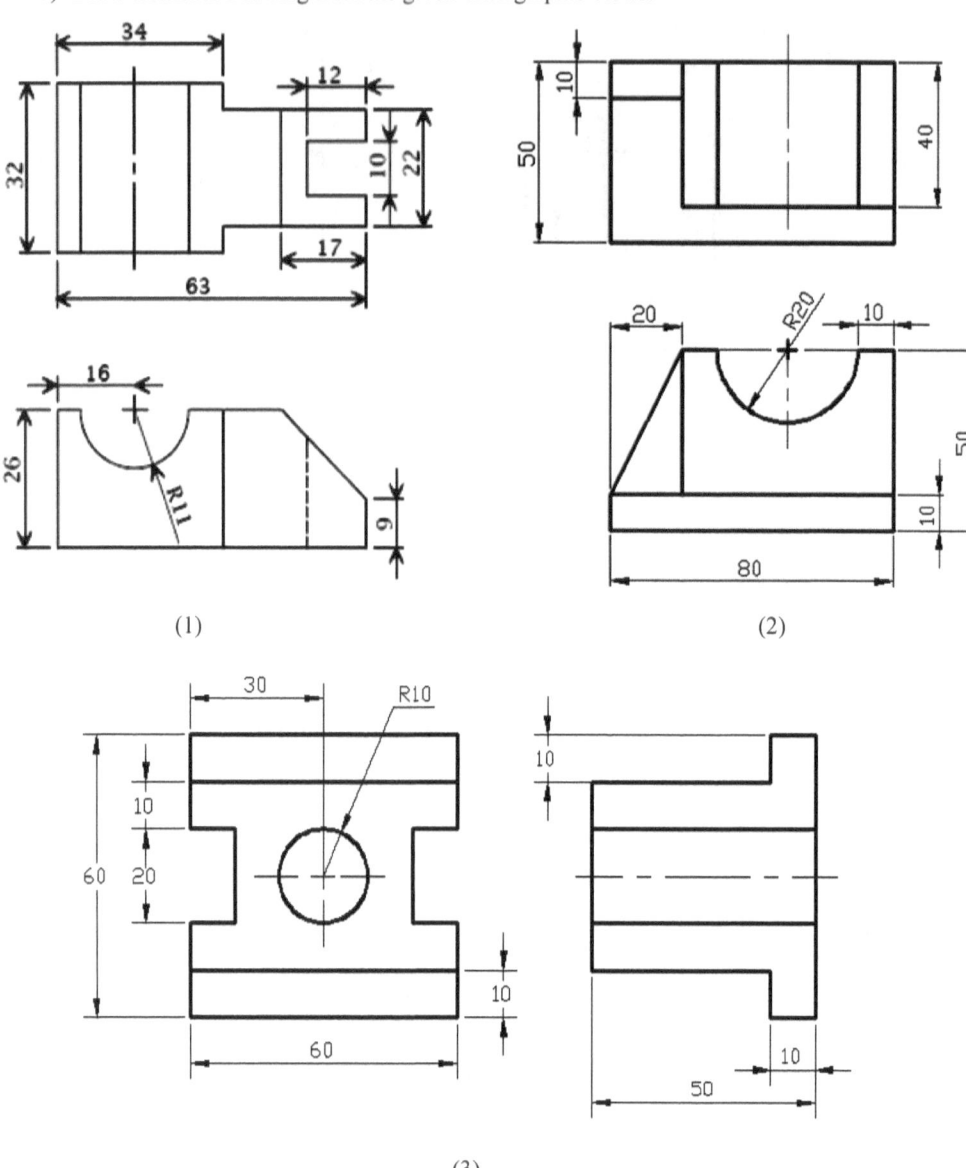

(1)

(2)

(3)

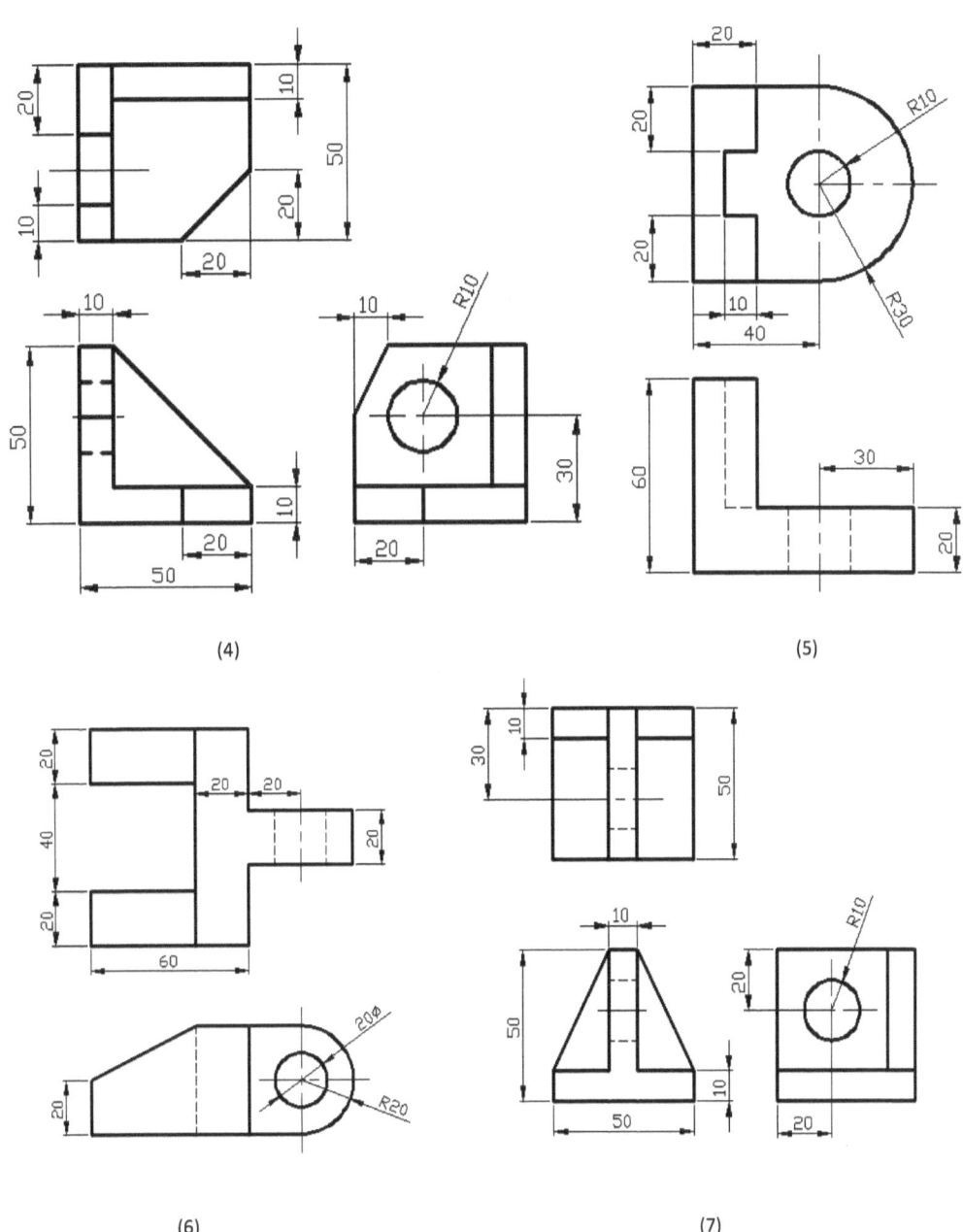

(4)

(5)

(6)

(7)

109

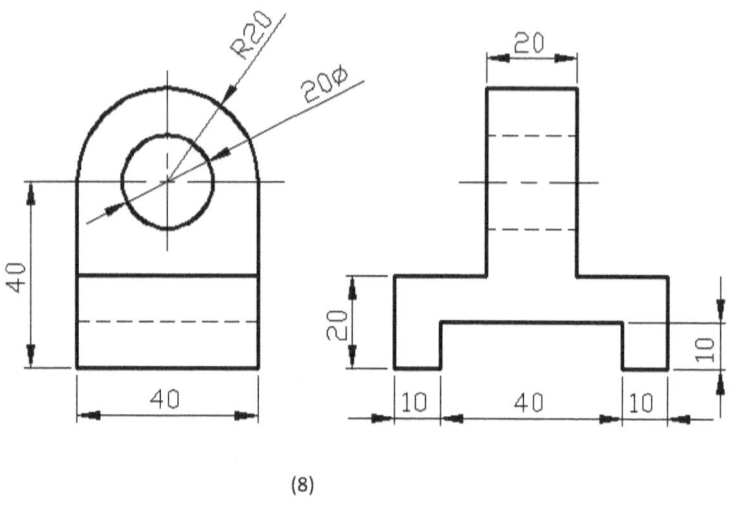

(8)

c) Draw oblique drawing of the following figures. Use 45o receding axis

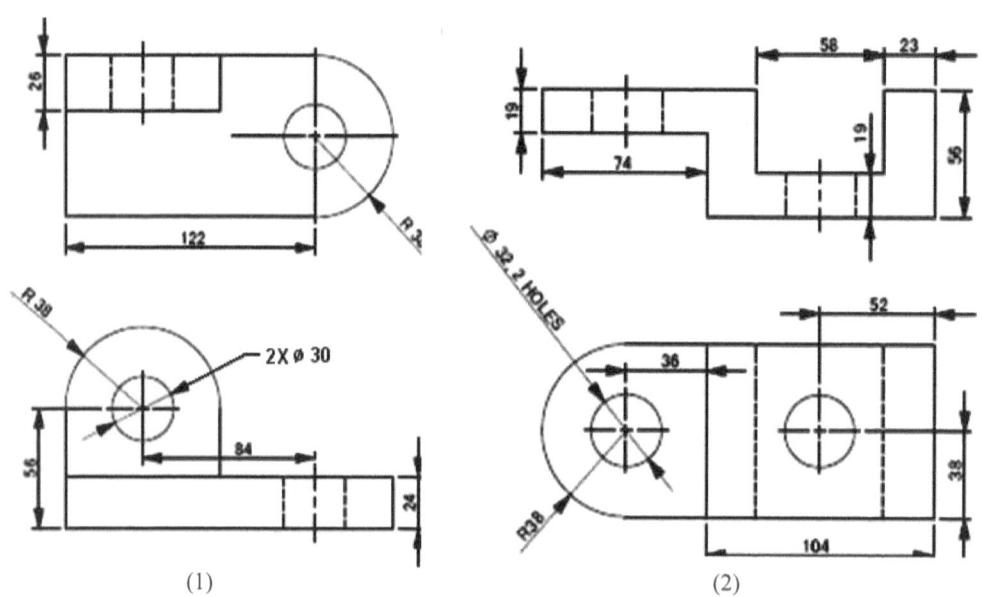

(1)   (2)

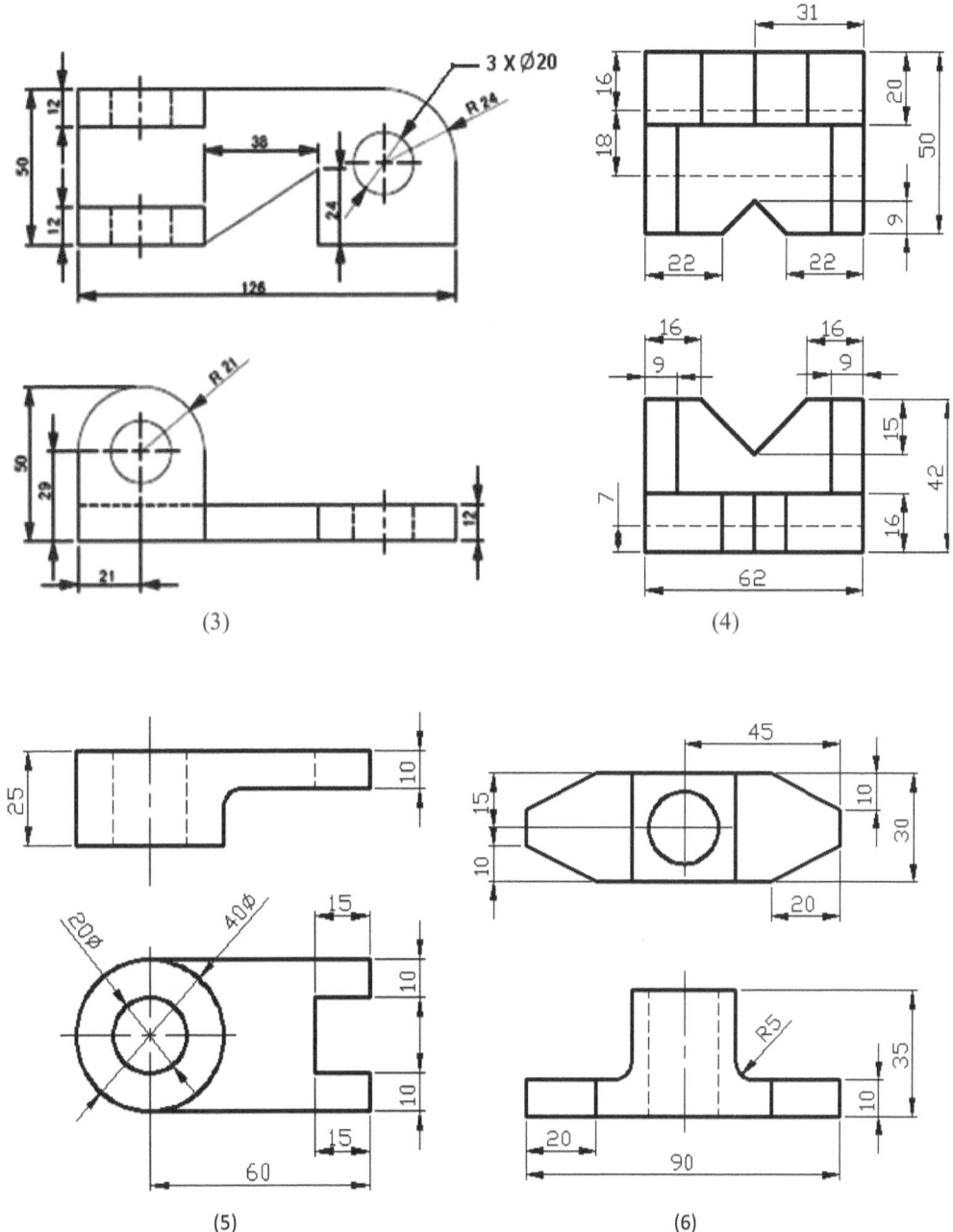

(3)

(4)

(5)

(6)

111

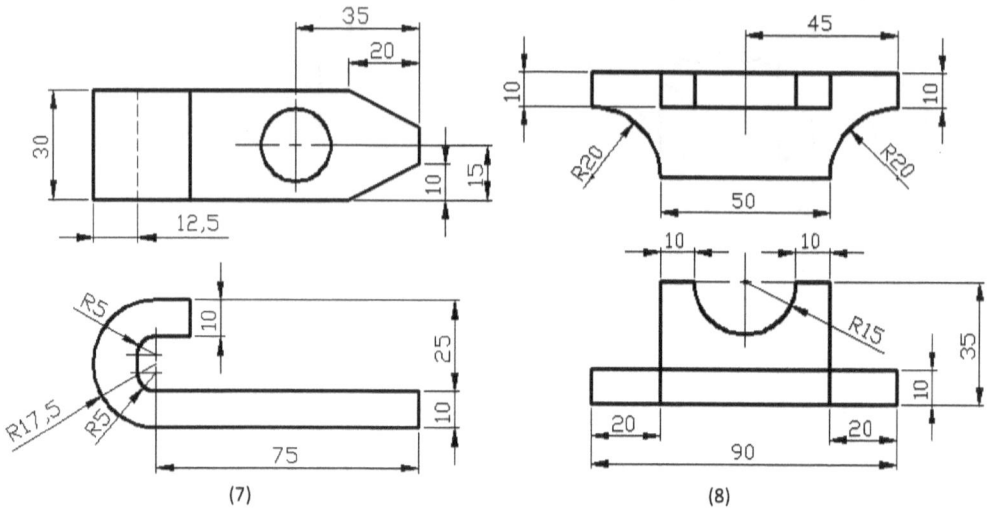

(7)   (8)

CHAPTER 5

AUXILIARY VIEWS

5.1 Introduction

There are times when the six principal views will not completely describe an object. This is especially true when there are inclined or oblique planes or features on an object. For these cases, a special orthographic view called an auxiliary view can be created.

An auxiliary view is an orthographic view that is projected into a plane that is not parallel to any of the principle planes. An auxiliary view is an orthographic view which is projected onto any plane other than the frontal, horizontal, or profile plane.

✓ An auxiliary view is not one of the six principal views.

✓ Inclined and oblique surfaces do not show true size in the standard views.

✓ Auxiliary views are often used to show inclined and oblique surfaces true size.

✓ Auxiliary views are orthographic views taken from a direction of sight other than top, front, right side, left side, bottom, or rear.

Fig. 5.1 A below shows three principal views of an object. Surface ABCD is an inclined plane and is therefore never seen in true size or shape in any of these views. In a multiview drawing, a true size and shape plane is shown only when the line of sight (LOS) used to create the view is perpendicular to the projection plane.

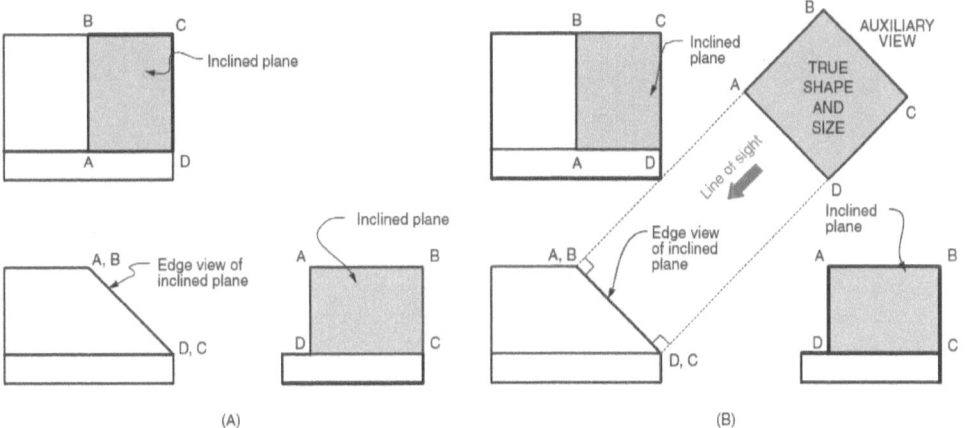

Fig. 5.1 Auxiliary view

113

To show the true size and shape of surface ABCD, an auxiliary view can be created by positioning a line of sight perpendicular to the inclined plane, then constructing the new view (Fig. 5.1B). In Fig. 5.2, the object is suspended in a glass box to show the six principal views, created by projecting the object onto the planes of the box. The box then is unfolded, resulting in the six principal views. However, when the six views are created, surface ABCD never appears true size and shape; it always appears either foreshortened or on edge.

Fig. 5.3 shows the object suspended inside a glass box, which has a special or *auxiliary plane* that is parallel to inclined surface ABCD. The line of sight required to create the auxiliary view is perpendicular to the new projection plane and to surface ABCD. The auxiliary plane is perpendicular to and hinged to the frontal plane, creating a *fold line* between the front view and the new auxiliary view.

In Fig. 5.4, the auxiliary glass box is unfolded, with the fold lines between the views shown as phantom lines. In the auxiliary view, surface ABCD is shown true size and shape and is located at distance M from the fold line.

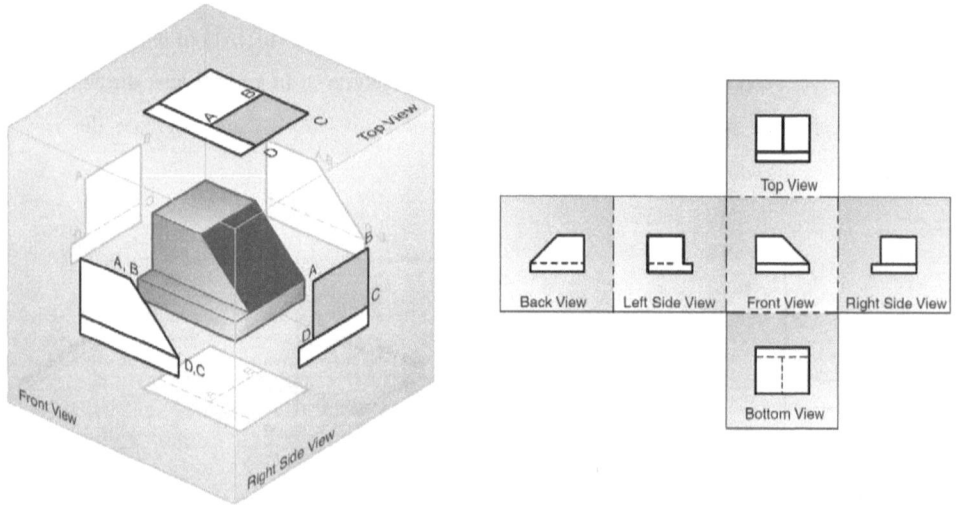

Fig. 5.2 Object in glass box and resulting six views when the box is unfolded

114

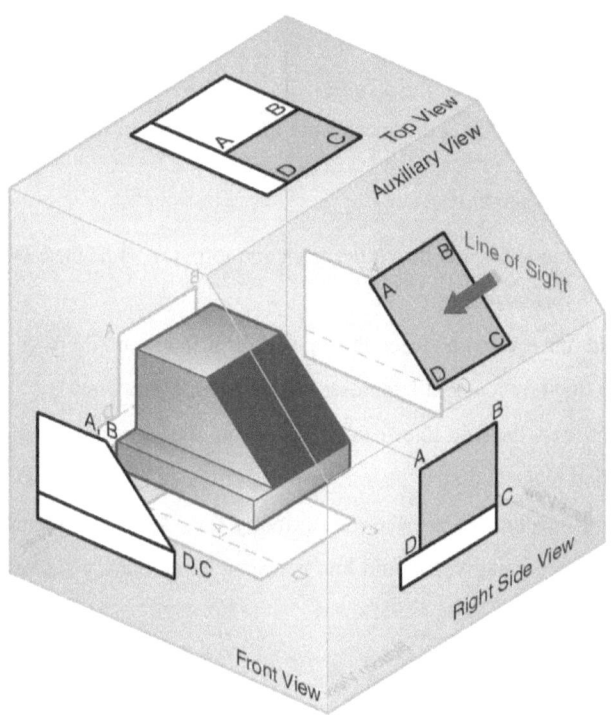

Fig. 5.3 Object in glass box with special auxiliary plane

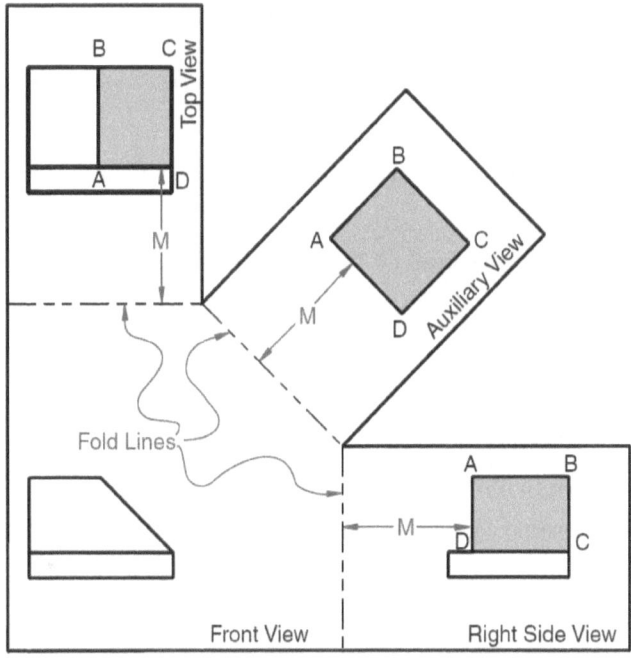

Fig. 5.4 Unfolding the glass box to create an auxiliary view of the inclined plane

Fold lines represent the edges of the "glass box". Orthographic lines are projected from adjacent views across fold lines. Object distances from fold lines are obtained from related views.

Fold line Labeling Conventions

✓ The fold line located between the front and top view is labeled **T-F**, where F means front and the T means top.

✓ The fold line located between the top and the primary auxiliary view is labeled **T-1**, where T is for the top view and 1 represents the first auxiliary view.

✓ Alternatively, the fold lines can be labeled by the projection planes. Since the horizontal projection plane contains the top view, the alternate labeling would be **H- F** and **H-1**.

✓ The fold line located between the primary (i.e., first) and secondary auxiliary views is labeled **1-2**. Similarly, the fold line between the secondary and tertiary auxiliary views is labeled **2-3**.

5.2 Auxiliary View Classifications

Auxiliary views are created by positioning a new line of sight relative to the object. It is possible to create any number of auxiliary views, including a new auxiliary view from an existing auxiliary view. Therefore, auxiliary views are first classified as: Primary, secondary, or tertiary.

✓ **A primary auxiliary view** is a single view projected from one of the six principal views.

✓ **A secondary auxiliary view** is a single view projected from a primary auxiliary view.

✓ **A tertiary auxiliary view** is a single view projected from a secondary or another tertiary auxiliary view.

Auxiliary views are also classified by the space dimension shown in true size in the primary auxiliary view.

✓ A **depth auxiliary** is an auxiliary view projected from the front view, and the depth dimension is shown true length.

✓ A **height auxiliary** view is an auxiliary view projected from the top view, and the height dimension is shown true length.

✓ A **width auxiliary** view is an auxiliary view projected from the profile view, and the width dimension is shown true length.

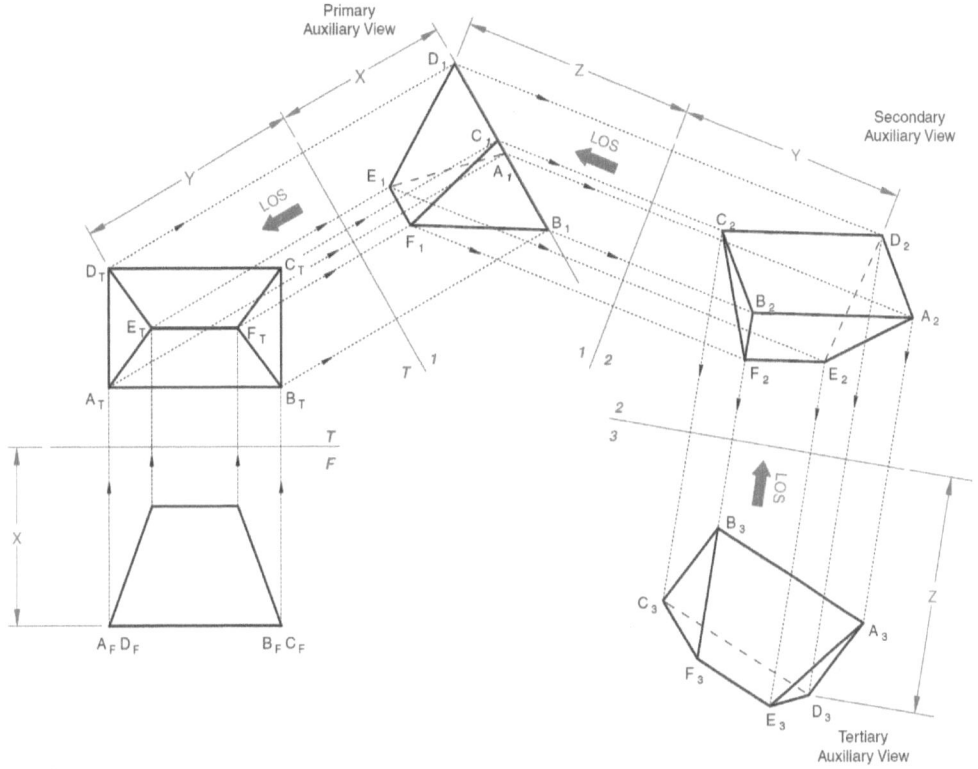

Fig. 5.5 Primary, secondary, and tertiary auxiliary views

The line of sight (LOS) determines the direction of the projection lines used in each auxiliary view

5.2.1 Depth Auxiliary View

A *depth auxiliary* view is projected from the front view, and the depth dimension is shown true length. Fig. 5.6 shows an auxiliary view that is projected from the front view of an object, using the fold-line method. Since plane ABCD is an inclined plane in the principal views, an auxiliary view is needed to create a true-size view of that plane. A depth auxiliary view of plane ABCD is created as described in the following steps.

Step 1: Given the front, top, and right side views, draw fold line F-1 using a phantom line parallel to the edge view of the inclined surface. Place line F-1 at any convenient distance from the front view.

117

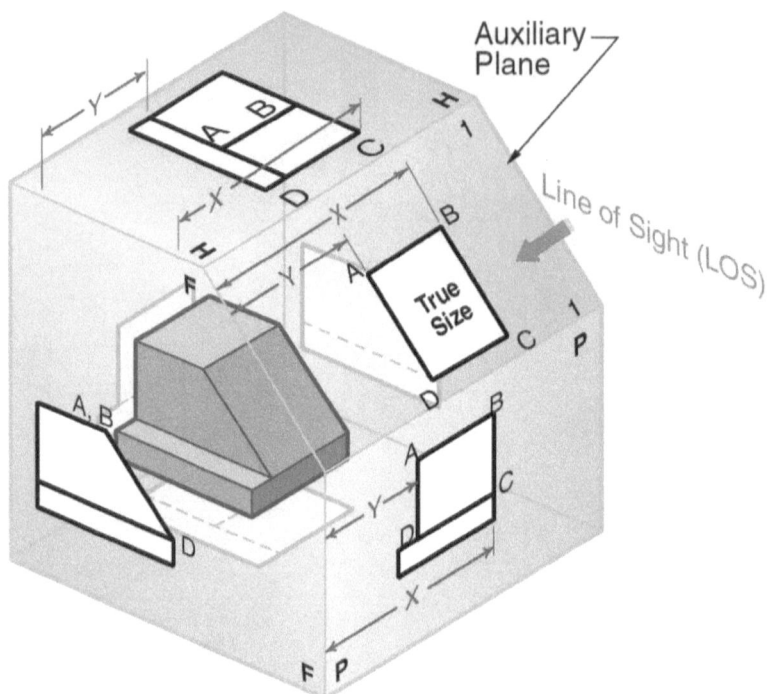

Fig. 5.6 Constructing a depth auxiliary view to determine the true size and shape of the inclined surface

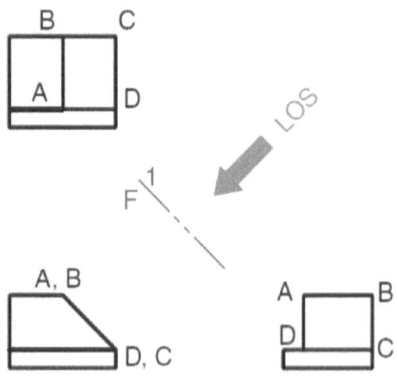

Step 1

Step 2: Draw fold line F-H between the front and top views. Line F-H should be perpendicular to the projectors between the views and at a distance X from the rear edge of the top view.

Draw fold line F-P between the front and the right side views, perpendicular to the projectors between the two views and at distance X from the rear edge of the right side view. The distance from fold line F-H to the top view must be equal to the distance from fold line F-P to the right side view. Draw parallel projectors between the principal views, using construction lines.

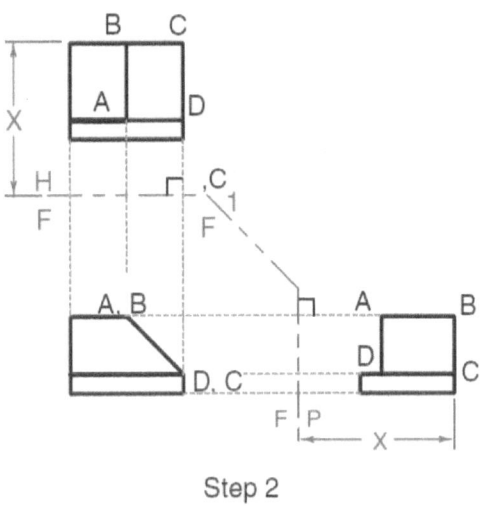

Step 2

Step 3: Project the length of the inclined surface from the top view to the auxiliary view, using construction lines. The projectors are perpendicular to the edge view and projected well into the auxiliary view from corners A, B and C, D.

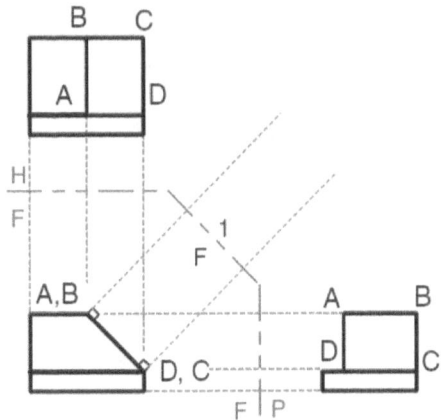

Step 3

119

Step 4: Transfer the depth of the inclined surface from the top view to the auxiliary view by first measuring the perpendicular distance from the fold line H-F to point C at the rear of the top view. This is distance X. Measure this same distance on the projectors on the projectors in the auxiliary view, measuring from fold line F-1. The measurement used to locate point C could have been taken from the profile view.

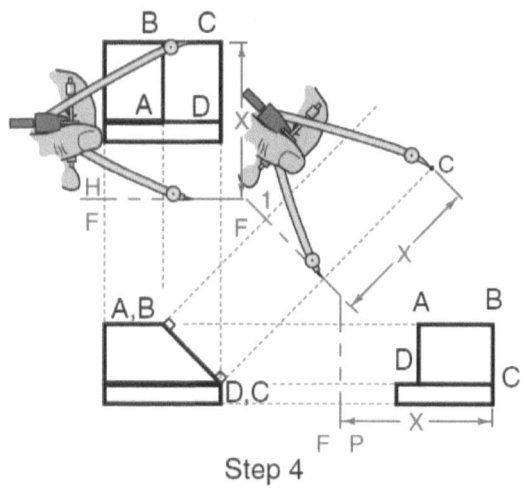

Step 4

Step 5: From point C in the auxiliary view, draw a line perpendicular to the projectors. Depth dimension Y is transferred from the top view by measuring the perpendicular distance from fold line H-F to point A (or D) in the top view and transferring that distance to the auxiliary view along the projectors perpendicular to fold line F-1. Draw a line at the transferred point A (or D) in the auxiliary view, perpendicular to the projectors.

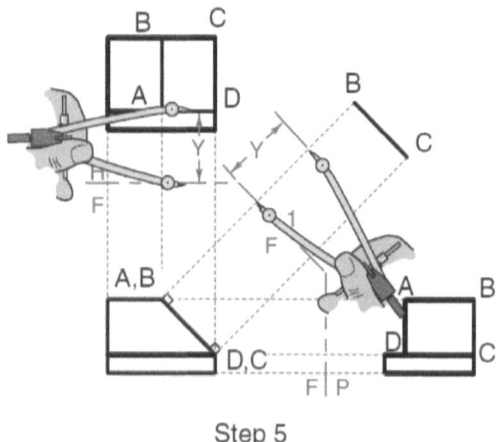

Step 5

Step 6: To complete the auxiliary view of the inclined surface, darken lines A-B and C-D.

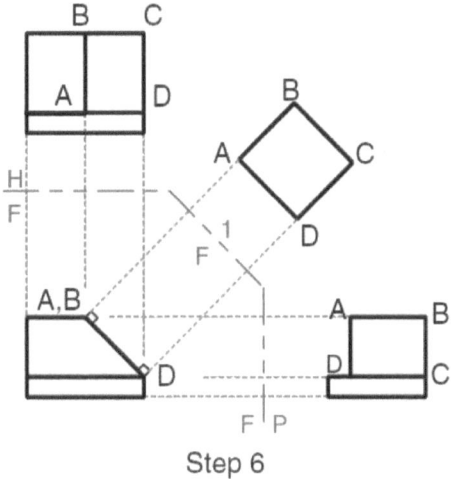

Step 6

5.2.2 Height Auxiliary View

A *height auxiliary view* is an auxiliary view projected from the top view, and the height dimension is shown true length. Fig. 5.7 shows an auxiliary view that is projected from the top view of an object, using the fold-line method. Since surface ABCD is an inclined plane in the principal views, an auxiliary view is needed to create a true-size view of that surface.

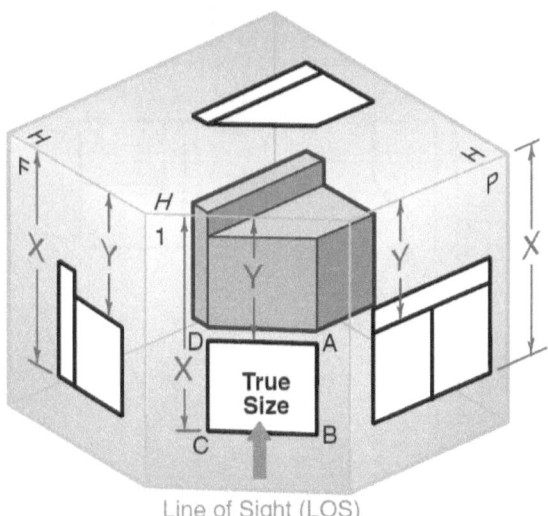

Fig. 5.7 Constructing a partial height auxiliary view of an inclined surface

A height auxiliary view is created as described in the following steps.

Step 1. Given the front, top, and right side views, draw fold line H–1 using a phantom line parallel to the edge view of the inclined surface. Place the line at any convenient distance from the top view.

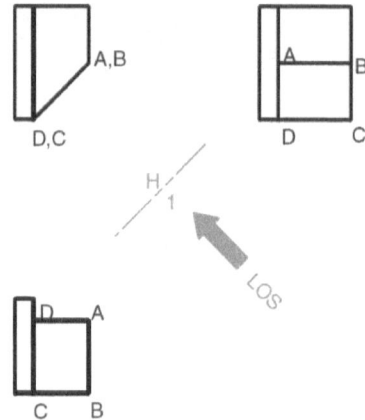

Step 2. Draw fold line H–F between the front and top views, perpendicular to the projectors between the views and at a distance X from the bottom edge of the front view. Draw fold line H–P between the top and right side views, perpendicular to the projectors between the views and at the same distance X from the rear edge of the right side view. The distance of fold line H–F to the front view must equal the distance from fold line H–P to the right side view. Draw parallel projectors between each principal view, using construction lines.

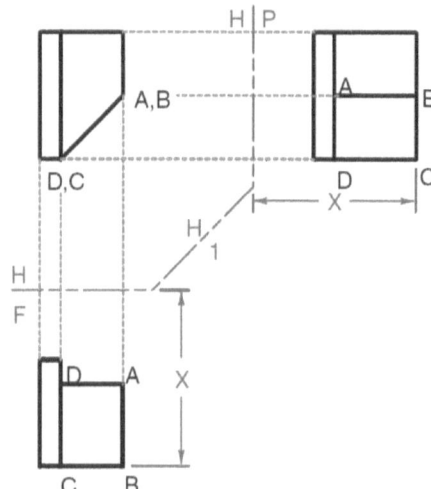

Step 3. Project the length of the inclined surface from the top view to the auxiliary view, using construction lines. The projectors are *perpendicular* to the edge view and projected well into the auxiliary view from corners A, B and D, C.

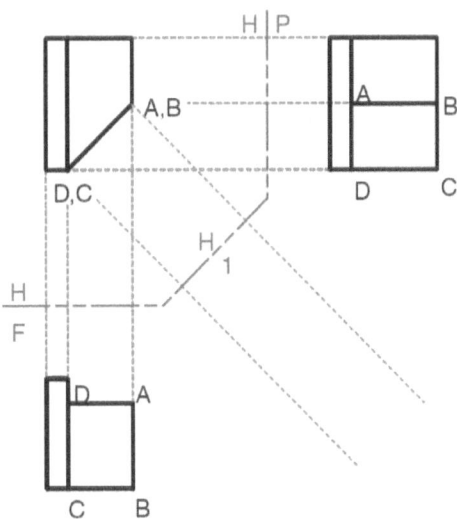

Step 4. Transfer the height of the inclined surface from the front view to the auxiliary view by first measuring the *perpendicular* distance from fold line H–F to the bottom edge of the front view. For this example, point C is measured at distance X from fold line H–F, and distance X is then measured along the projectors *perpendicular* to fold line H–1. From point C in the auxiliary view, draw a line perpendicular to the projectors.

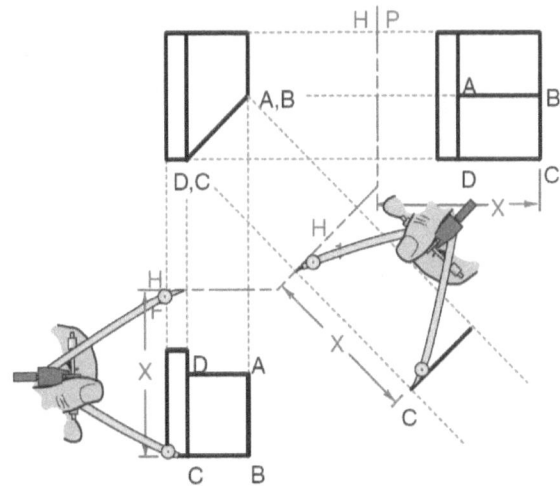

Step 5. Height dimension Y then is transferred from the front view in a similar manner, measuring the *perpendicular* distance from fold line H–F to point A (or D) of the front view and transferring this distance to the auxiliary view, measuring along the projectors *perpendicular* to fold line H–1. From the transferred point A *in* the auxiliary view, draw a line perpendicular to the projectors.

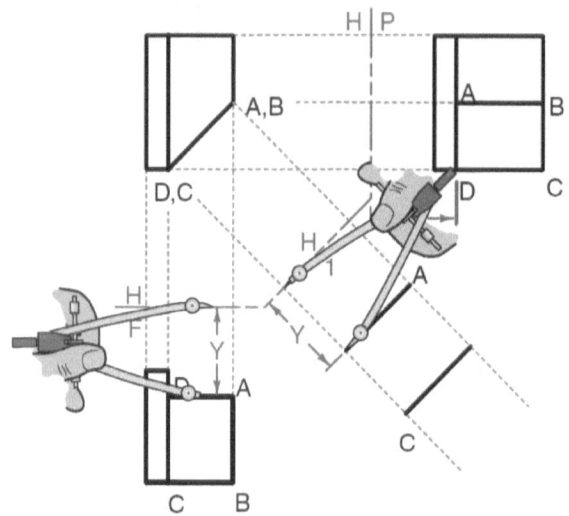

Step 6. Darken lines A–B and D–C to show the true size of the inclined surface and to complete the partial auxiliary view.

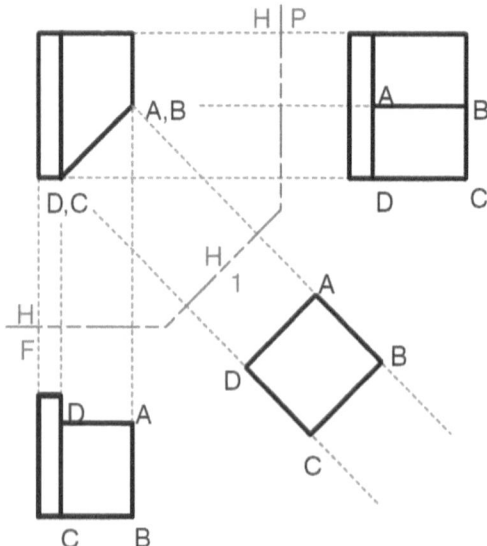

5.2.3 Width Auxiliary View

A *width auxiliary view* is an auxiliary view projected from the profile view, and the width dimension is shown true length. Figure 5.8 shows an auxiliary view that is projected from the profile view of an object, using the fold line method. Since plane ABCD is an inclined plane in the principal views, an auxiliary view is needed to create a true-size view of the plane.

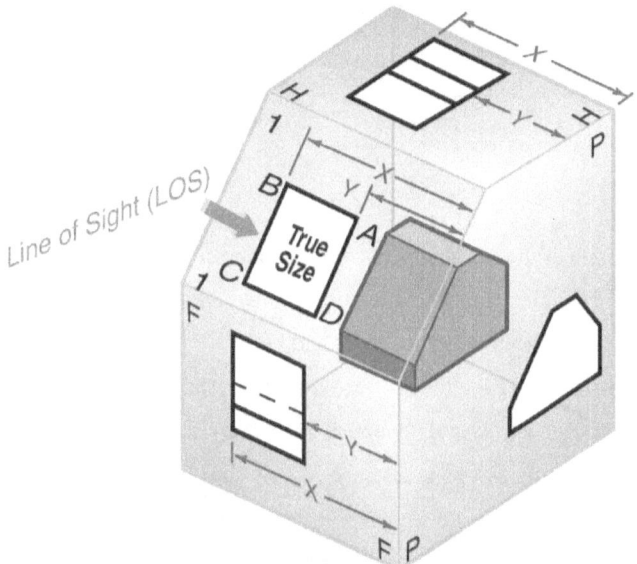

Fig. 5.8 Constructing a partial width auxiliary view of an inclined surface

A width auxiliary view is created as described in the following steps.

Step 1. Given the front, top, and left side views, draw fold line P–1 using a phantom line parallel to the edge view of the inclined surface. Place the line at any convenient distance from the profile view.

Step 2. Draw fold line F–P between the front and profile views, perpendicular to the projectors between the views and at a distance X from the left edge of the front view. Draw fold line H–P between the top and profile views, perpendicular to the projectors between the views, and at a distance X from the rear edge of the top view. The distance from fold line H–P to the top view must equal the distance from fold line F–P to the front view. Draw parallel projectors between each view, using construction lines.

125

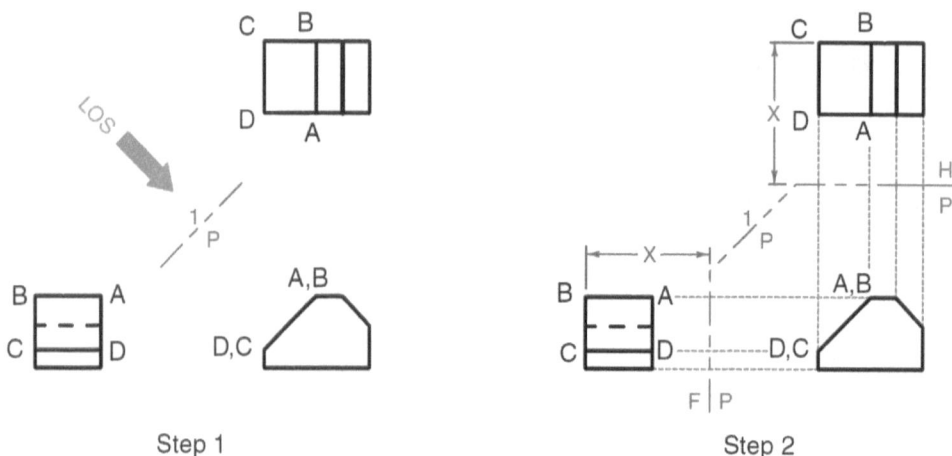

Step 1

Step 2

Step 3. Project the length of the inclined surface from the profile view to the auxiliary view, using construction lines. The projectors are *perpendicular* to the edge view and projected well into the auxiliary view from corners A, B and D, C.

Step 4. Transfer the width of the inclined surface from the front view by first measuring the *perpendicular* distance from fold line P–F to the left side of the front view. For this example, point B is measured at distance X from fold line P–F and is then transferred to the auxiliary view along the projectors *perpendicular* to fold line P–1. From point B in the auxiliary view, draw a line perpendicular to the projectors

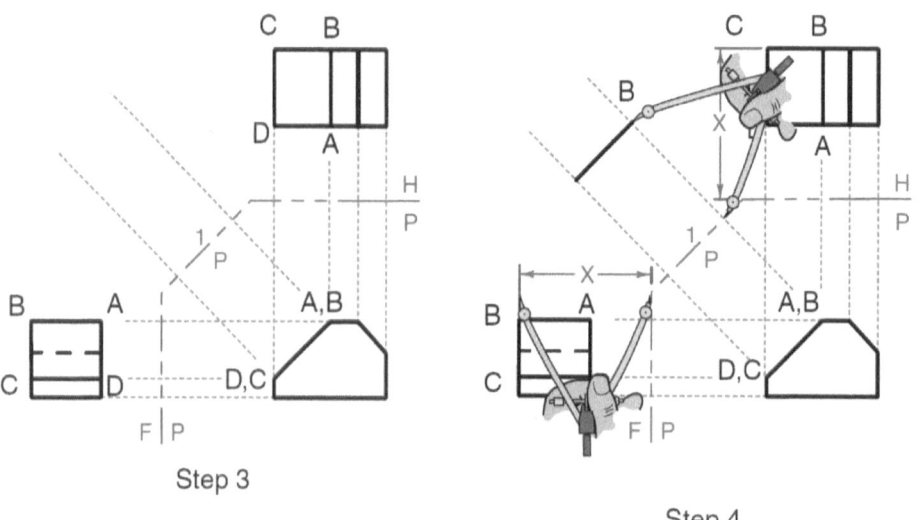

Step 3

Step 4

Step 5. Width dimension Y is then transferred from the front view in a similar manner, measuring the *perpendicular* distance from fold line P–F to point A (or D) of the front view and transferring this distance to the auxiliary view along the projectors *perpendicular* to fold line P–1. From the transferred point A in the auxiliary view, draw a line perpendicular to the projectors.

Step 6. Darken lines A–B and C–D to show the true size of the inclined surface to complete the partial auxiliary view.

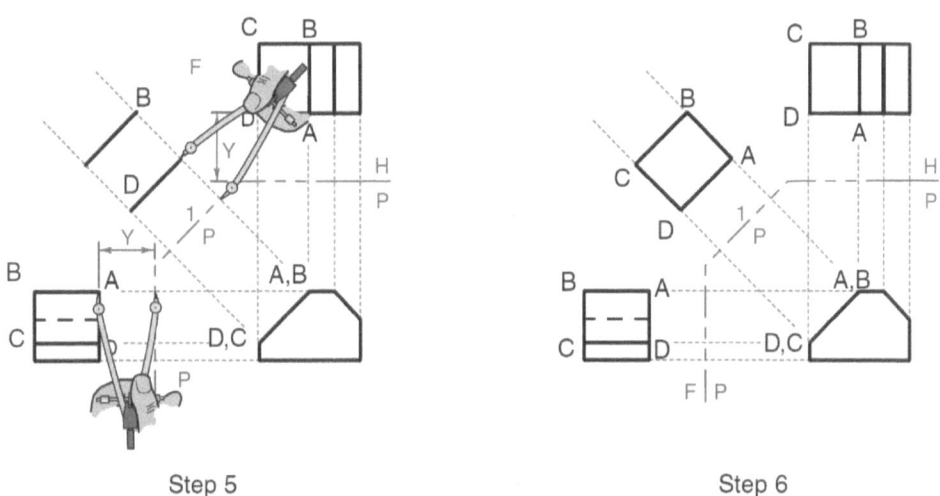

Step 5   Step 6

5.2.4   Partial Auxiliary Views

In auxiliary views, it is normal practice not to project hidden features or other surfaces that are not part of the inclined surface. When only the details of the inclined surface are projected and drawn in the auxiliary view, the view is called a **partial auxiliary** view. A partial auxiliary view saves time and produces a drawing that is much more readable. The figure below shows a partial and full auxiliary view of the same object. The full auxiliary view is harder to draw, read, and visualize. In this example, some of the holes would have to be drawn as ellipses in the full auxiliary view. A break line can be used to indicate that the view is a partial view.

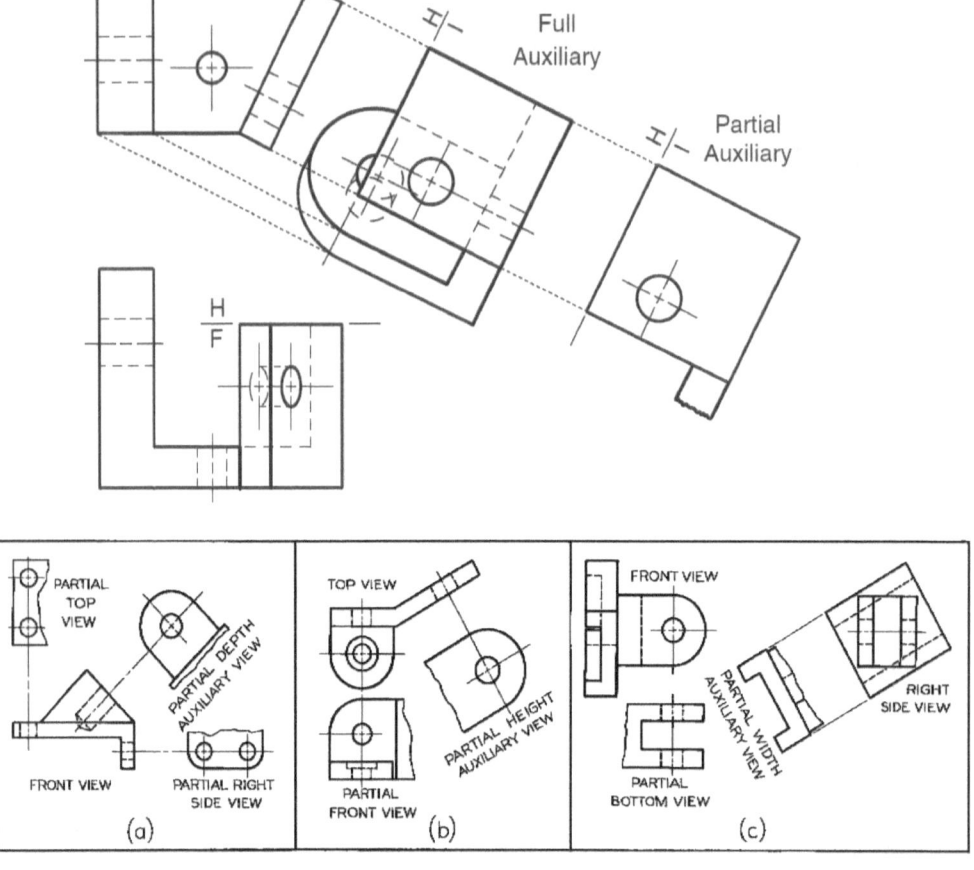

Fig. 5.9 A full auxiliary view, including hidden lines (above), and a partial auxiliary view with no hidden lines

5.2.5 Half Auxiliary Views

Symmetrical objects can be represented as a **half auxiliary view;** that is, only half of the object is drawn.

128

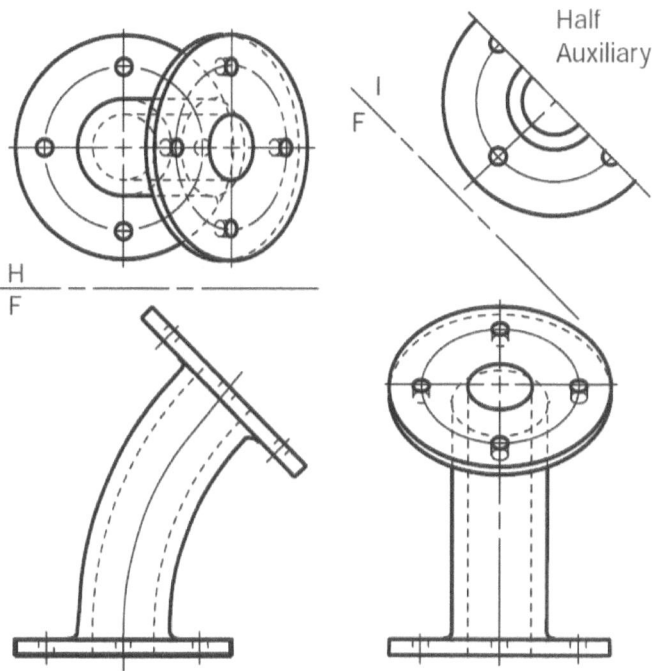

Fig. 5.10 Half auxiliary view of a symmetrical feature

5.2.6 Constructing a Curve in an Auxiliary View

Fig. 5.11 below shows a cylindrical part that is cut by an inclined plane. The resulting surface is an ellipse that can be shown true size and shape with auxiliary view. The process for drawing curves in an auxiliary view is described in the following steps.

Step 1: In the right side view, locate a reference plane at the vertical center of the cylinder. The reference plane will be coincident to the axis of the cylinder and is therefore shown as an edge view in the right side view. The reference plane is located in the center so that all dimensions can be located on either side of the corresponding reference plane in the auxiliary view.

Step 2: Locate the edge view of the reference plane in the auxiliary view by drawing a line parallel to the edge view of the ellipse and at any convenient distance from that edge. The reference plane will coincide with the location of the major axis of the ellipse. The location of the reference plane should leave sufficient room for the auxiliary view to be plotted without running into any of the multiviews.

129

Step 3: Plot points along the circumference of the circle in the right side view, and project these points onto the edge view of the ellipse in the front view. Number the points to assist in plotting the corresponding points in the auxiliary view.

Step 4: Project the points from the ellipse edge view in the front view through the reference plane in the auxiliary view. The projectors should be perpendicular to the edge view and the reference plane. The projector from the point for the center line of the cylinder in the front view coincides with the minor axis of the ellipse in the auxiliary view. Measure and transfer the depth dimensions from the right side view to the projectors in the auxiliary view.

Step 5: Using a French curve, connect the points to create the true size and shape of the curved surface.

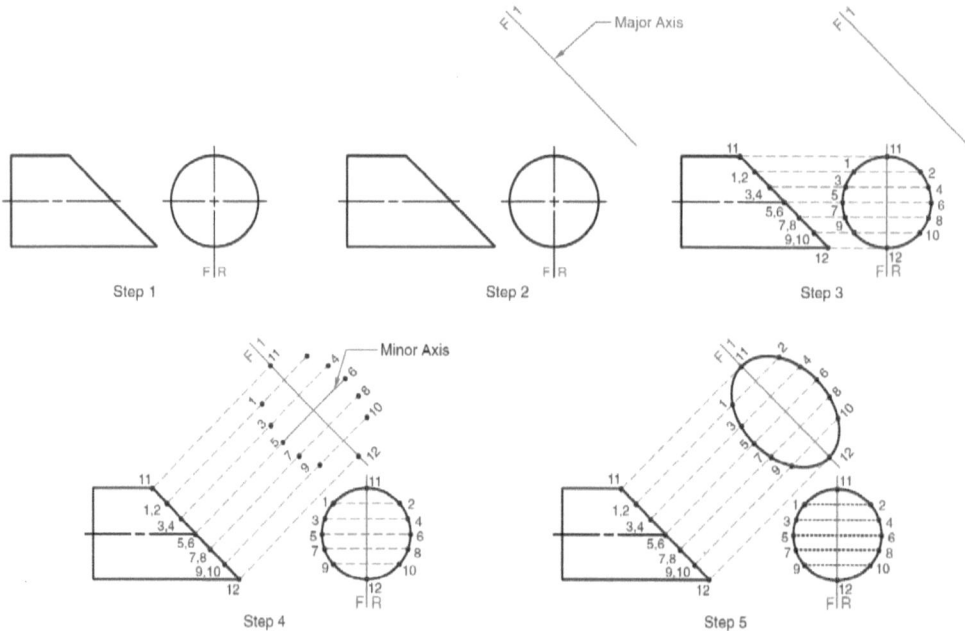

Fig. 5.11 Constructing a Curve in an Auxiliary View

130

5.2.7 Secondary (Oblique) Auxiliary Views

Frequently an object will have an inclined face that is not perpendicular to any of the principal planes of projection. In such cases it is necessary to draw a primary auxiliary view and a secondary auxiliary or oblique view.

The primary auxiliary view is constructed by projecting the figure on a primary auxiliary plane that is perpendicular to the inclined surface and one of the principal planes. This plane may be at any convenient location. The inclined face appears as a straight line in the primary auxiliary view. Using this view as a regular view, the secondary auxiliary view may be projected on a plane parallel to the inclined face.

Successive auxiliary views can be used to draw an oblique surface in true size and shape. The first step is to construct a new view from one of the principal views, parallel to a true-length line of the oblique plane. In this new view, the oblique surface will be an edge. A secondary auxiliary view is then created, perpendicular to projectors from the edge view of the oblique surface, and the secondary view shows the true size and shape of the surface. The following steps describe how to create a true-size view of the oblique surface in Fig. 5.12.

Step 1. For the first auxiliary view, place the line of sight parallel to a true-length line of the oblique surface, in one of the principal views. For this example, side A–B of the oblique triangular surface ABC is a true-length line in the top view. Draw a projector from point B, parallel to line A–B. Draw a line parallel to this projector, from point C. Draw reference plane H–1 perpendicular to these projectors. Place another reference plane H–F between the front and top views, perpendicular to the projectors between the two views. In the front view, measure the perpendicular distances from reference line H–F to points A and C. Transfer these measurements to the auxiliary view, measuring along the projectors, from reference line H–1. This will produce an edge view of the oblique surface, labeled B, A–C.

Step 2. Create a secondary auxiliary view by projecting lines from points A, B, and C, perpendicular to the edge view of the surface. Draw a reference line 1–2, perpendicular to these projectors. Measure the perpendicular distances from reference line H–1 to points B, A, and C in the top view. Transfer these measurements to the secondary auxiliary view, measuring along the projectors, from reference line 1–2. Darken lines A–B, B–C, and C–A, to produce a true-size view of the oblique surface ABC.

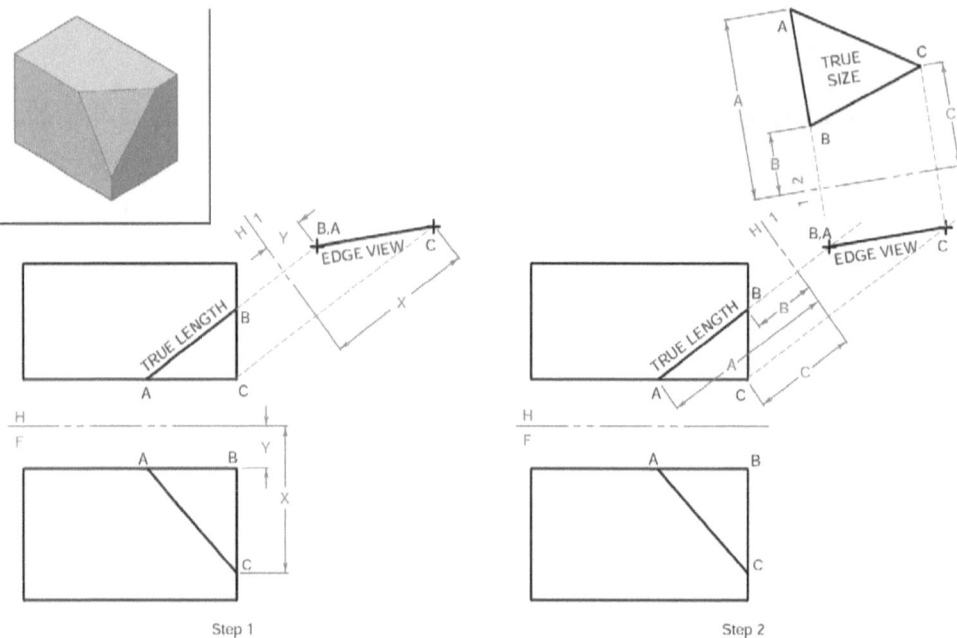

Step 1   Step 2

Fig. 5.12 Constructing successive auxiliary views to determine the true size of an oblique surface

5.3 Problems

1. Draw the two given views, then create a partial auxiliary view of the inclined surface. Each grid space equals 1.5 cm.

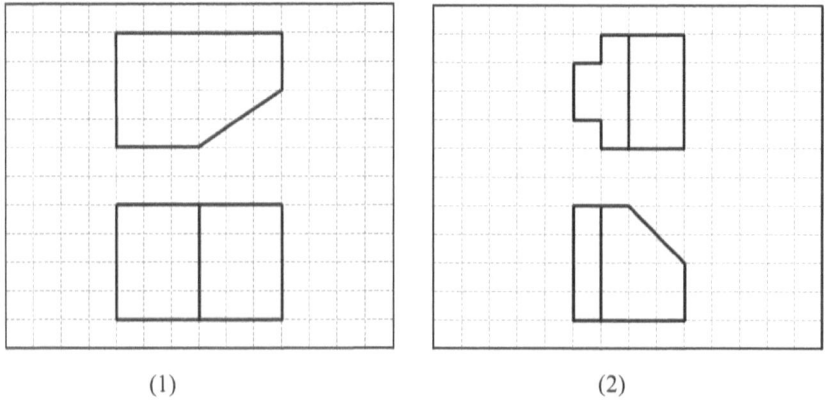

(1)   (2)

132

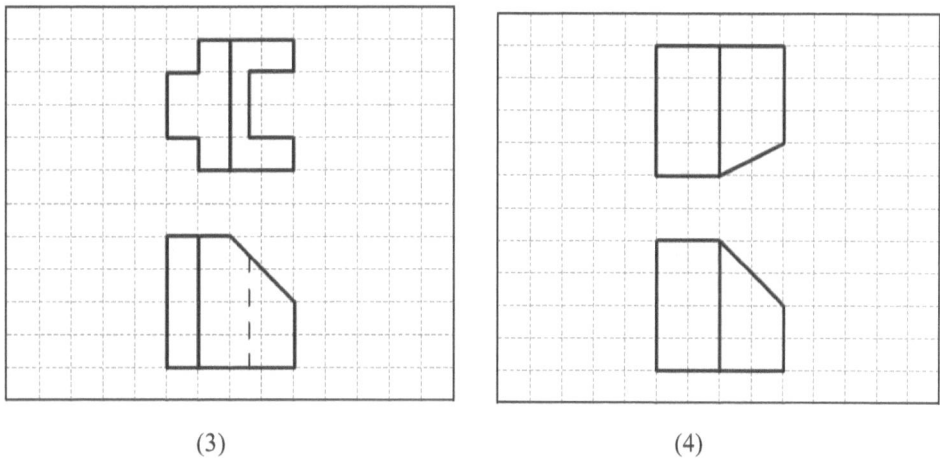

(3)   (4)

2. Make the *partial auxiliary view* of the objects whose front and top views are given below.

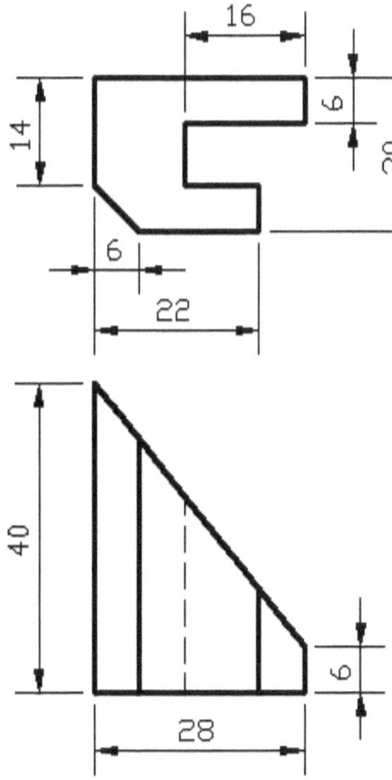

(1) 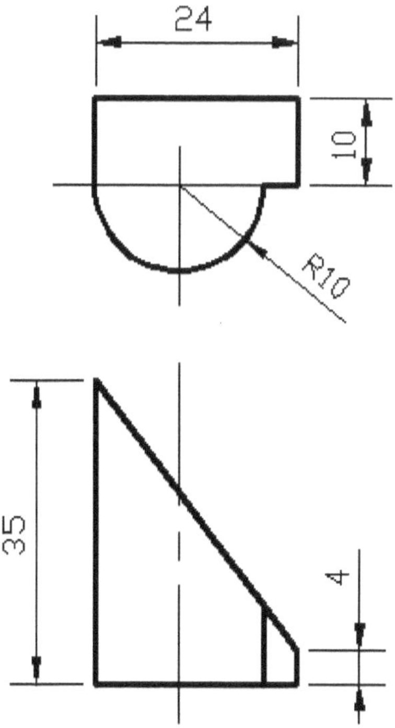   (2)

3. Make the **Complete auxiliary view** of the objects whose front and top views are given below

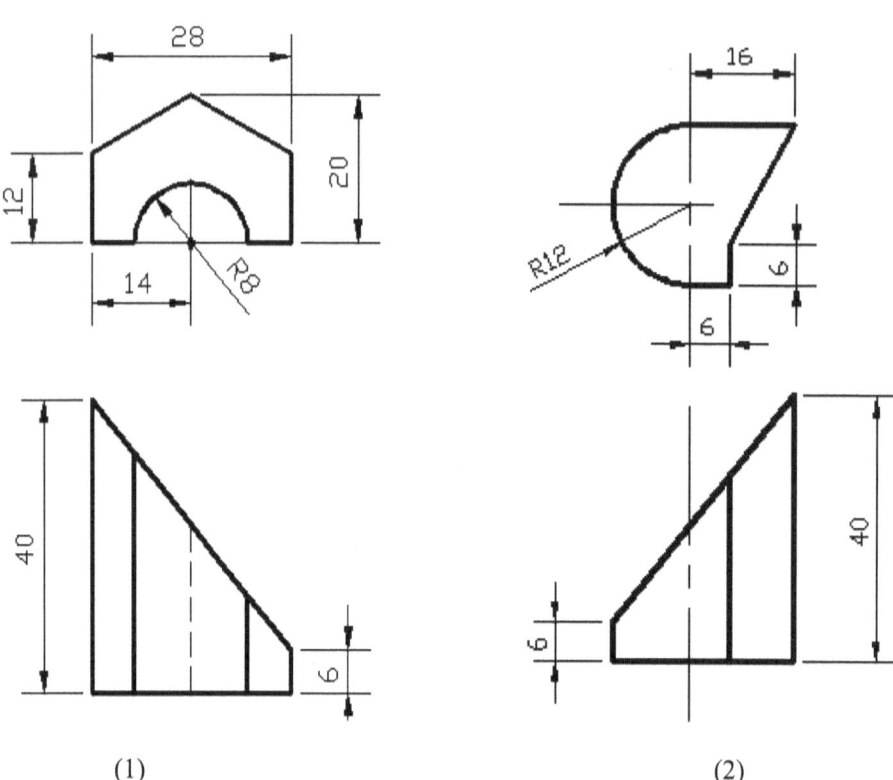

(1)   (2)

CHAPTER 6
SECTIONAL VIEWS

6.1 Introduction

In an orthographic projection drawing, outlines and edges of an object are usually depicted with continuous lines and internal details are normally illustrated by using hidden lines. When dealing with complex objects, there may be many hidden lines and these hidden lines may become very confusing. The technique called *section views* is used to 'cut sections' across the object to show internal details.

S*ection views* is also used to improve the visualization of new designs, clarify multiview drawings, and facilitate the dimensioning of drawings. For example, orthographic drawings of complicated mechanical parts can be very difficult to visualize and dimension. Section views can be used to reveal interior features of an object that are not easily represented using hidden lines. Architectural drawings use section views to reveal the interior details of walls, ceilings, and floors. Three-dimensional geometric models created with CAD can be sectioned to reveal interior features and assist in the design of complex systems.

Section views use a technique that is based on *passing an imaginary cutting plane through a part to reveal interior features*. Creating section views requires visualization skills and adherence to strict standards and conventional practices. This chapter will explain the theory for creating section views, the important conventional practices used for section views, and will show examples of standard practices.

Sectional views are multiview technical drawings that contain special views of a part or parts, views that reveal interior features of the part. *A primary reason for creating a sectional view is the elimination of hidden lines so that a drawing can be more easily understood or visualized*. Sectional views use a technique that is based on **passing an imaginary cutting plane** through a part **to reveal interior features**.

Sectional views are usually produced

    a)   To clarify details of the object,

    b)   To illustrate internal features clearly,

    c)   To reduce the number of hidden-detail lines,

d) To facilitate the dimensioning of internal features,

e) To show the shape of the cross-section,

f) To show clearly the relative positions of parts forming an assembly.

Fig. 6.1 below shows a regular multiview drawing and a sectioned multiview drawing of the same part in the front view; the hidden features can be seen after sectioning.

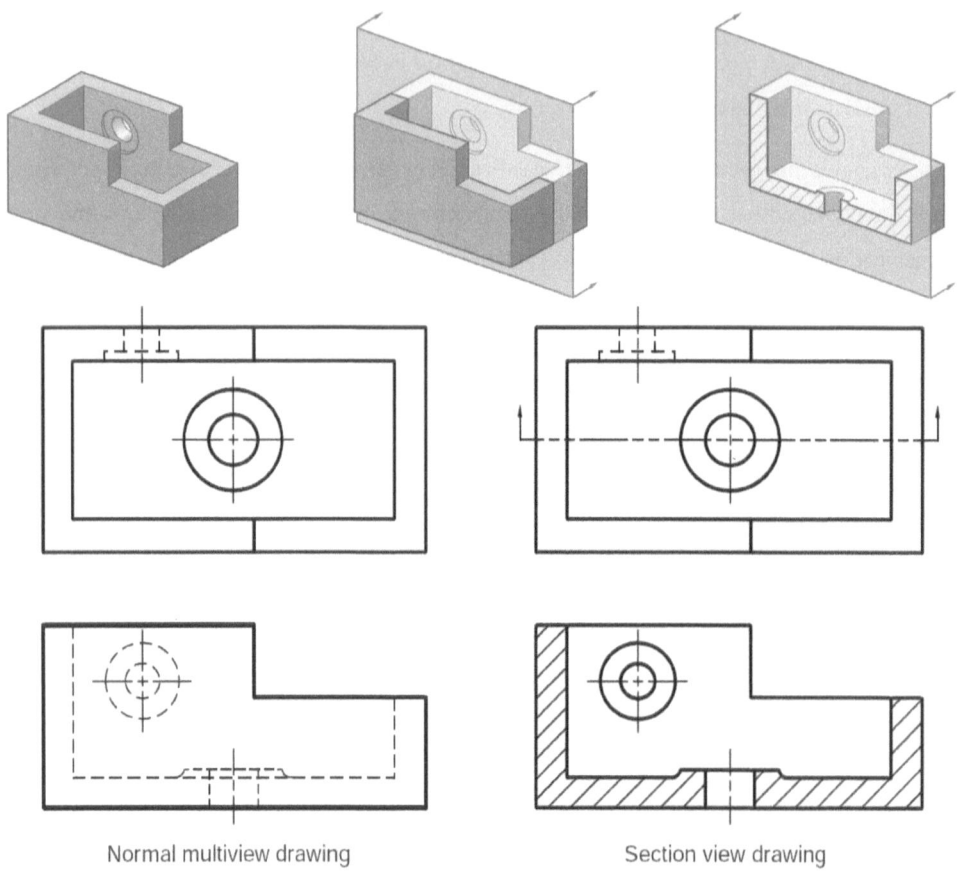

Normal multiview drawing   Section view drawing

Fig. 6.1 Sectional view reveals hidden features

6.2 Cutting Plane

In order to show the internal features without excessive use of hidden-detail lines, the object is imagined to be cut along a plane called a ***cutting plane***. The cut portion nearer to the observer is removed and the remaining part is shown as a sectional view. The surfaces in

136

section can be imagined to be cut along the cutting plane with an imaginary tool and imaginary cutting marks are represented by thin equidistant **hatching lines** as shown in Fig. 6.2. Hidden portion of the object behind the section is generally omitted and not shown by hidden lines unless it is very essential for further clarification. The imaginary cutting plane is controlled by the designer and can (1) go completely through the object (full section), (2) go halfway through the object (half section), (3) be bent to go through features that are not aligned (offset section), or (4) go through part of the object (broken-out section).

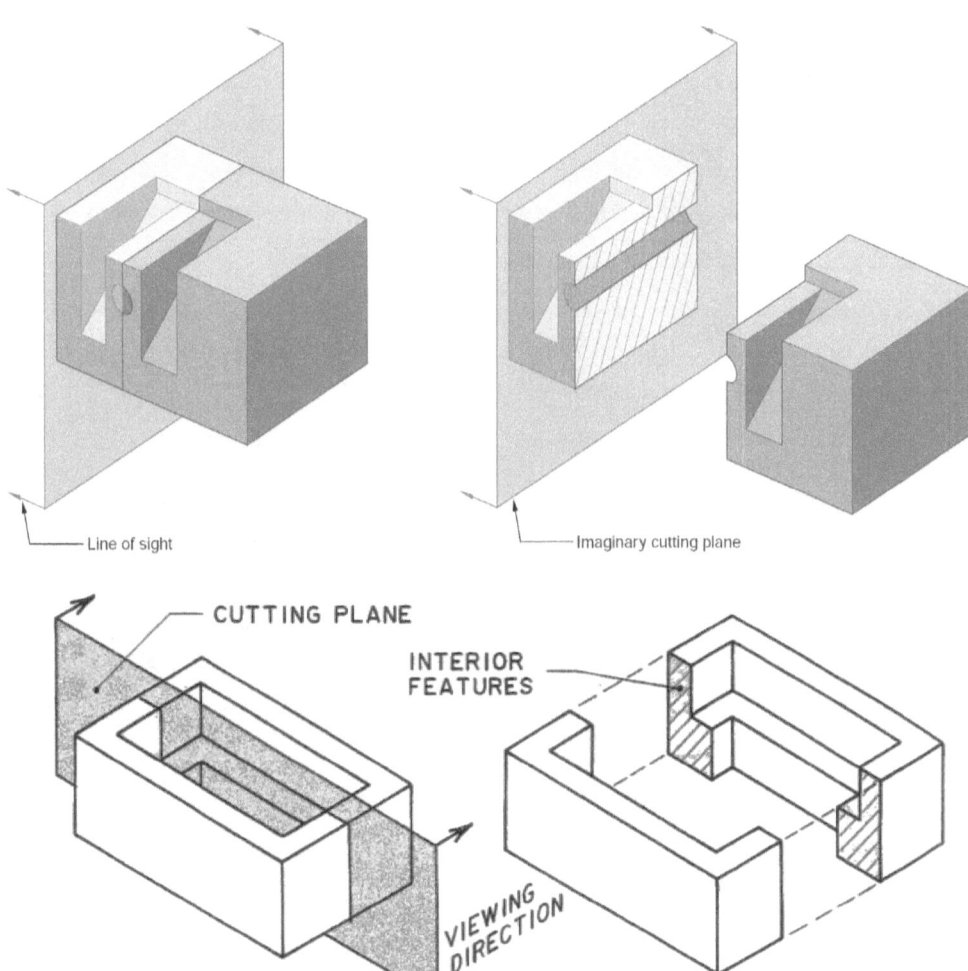

Fig. 6.2 Cutting planes. Imaginary cutting planes used to create section views are passed through the object to reveal interior features.

137

6.3 Cutting Plane Lines

Cutting plane lines, which show where the cutting plane passes through the object, represent the edge view of the cutting plane and are drawn in the view(s) adjacent to the section view. Cutting plane lines are thick (0.6mm) dashed lines that extend past the edge of the object 6 mm and have line segments at each end drawn at 90 degrees and terminated with arrows. The arrows represent the direction of line of sight for the section view. Two types of lines are acceptable for cutting plane lines in multiview drawings.

Line B–B (Fig. 6.3) is composed by alternating one long and two short dashes. The length of the long dashes varies according to the size of the drawing and is approximately 20 to 40 mm. For a very large section view drawing, the long dashes are made very long to save drawing time. The short dashes are approximately 3 mm long. The open space between the lines is approximately 1.5 mm. Capital letters are placed at each end of the cutting plane line, for clarity or for differentiating between cutting planes when more than one is used on a drawing.

The second method used for cutting plane lines, shown by line C–C in Fig. 6.3, is composed of equal-length dashed lines. Each dash is approximately 6 mm long, with a 1.5 mm space in between. Standard cutting plane linestyles are thick lines terminated with arrows.

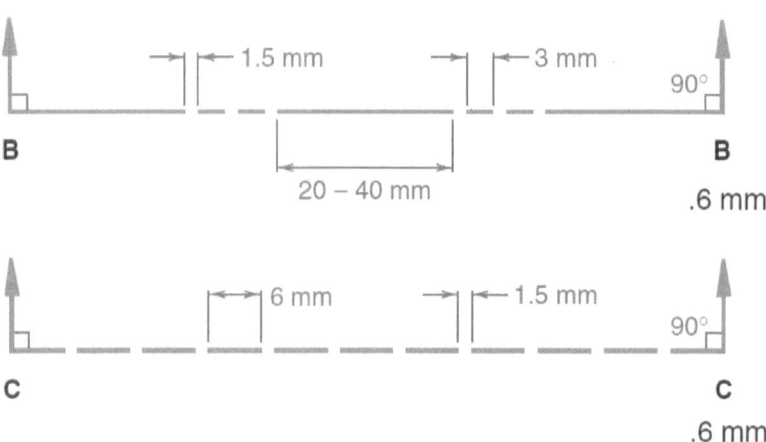

Fig. 6.3 Standard cutting plane linestyles

6.3.1 Placement of Cutting Plane Lines

The cutting plane line is placed in the view where the cutting plane appears on edge. If the cutting plane appears as an edge in the top view and is normal in the front view, it is a **frontal cutting plane.** The front half of the object is "removed", and the front view is drawn in section.

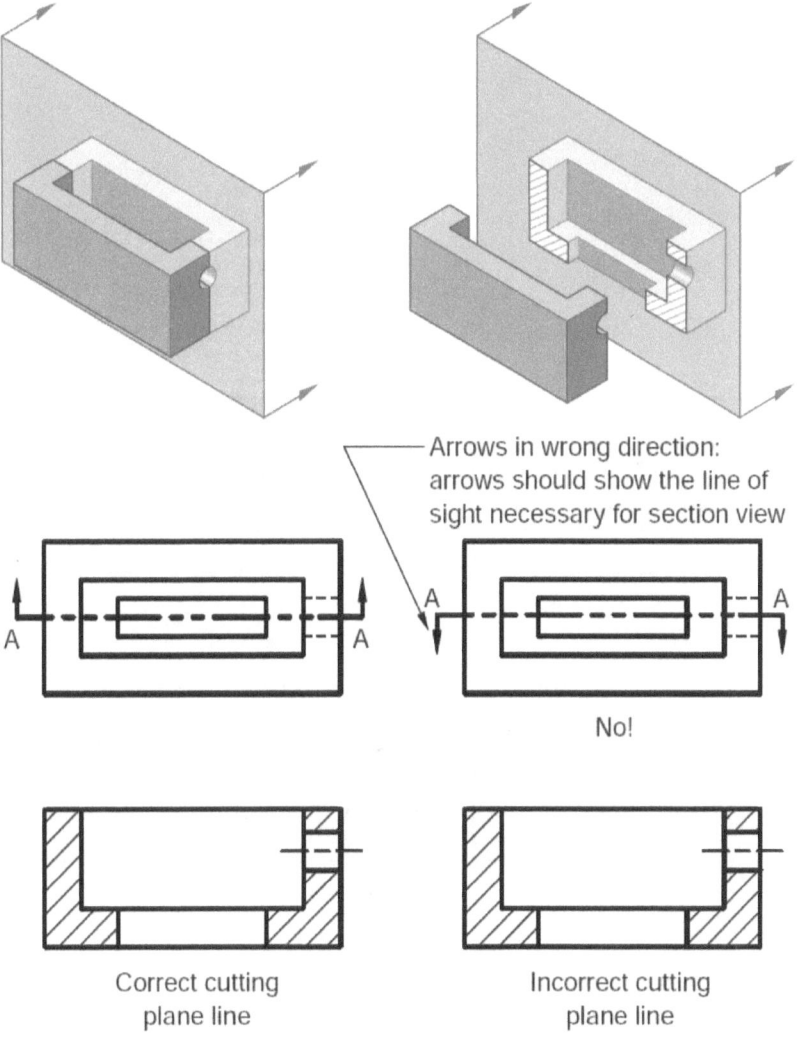

Fig. 6.4 Placement of cutting plane lines in frontal section view

A horizontal section view is one where the cutting plane is on edge in the front view and the top view is sectioned. If the cutting plane appears as an edge in the front view and is normal in the top view, it is a **horizontal cutting plane.** The top half of the object is "removed", and the top view is drawn in section.

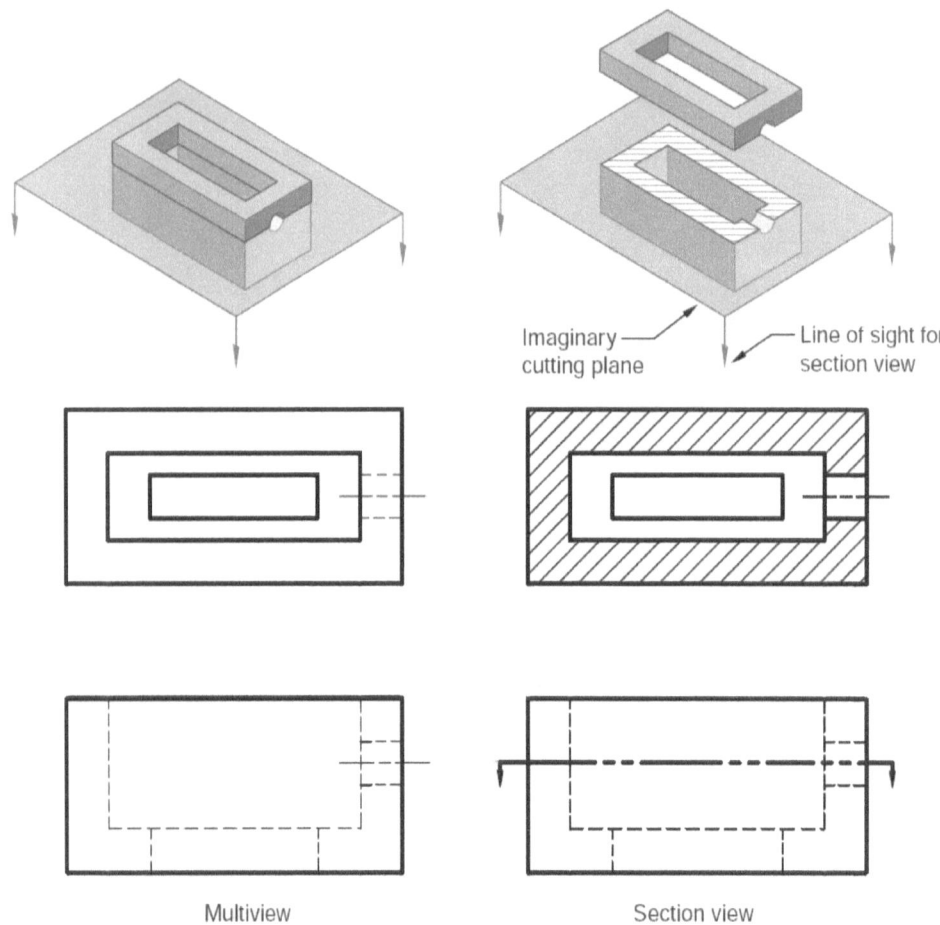

Fig. 6.5 Horizontal section view

If the cutting plane appears as an edge in the top and front views and is normal in the profile view, it is a **profile cutting plane.** The left (or right) half of the object is "removed", and the left (or right) side view is drawn in section. A profile section view is one in which the cutting plane is on edge in the front and top views and the profile view is sectioned.

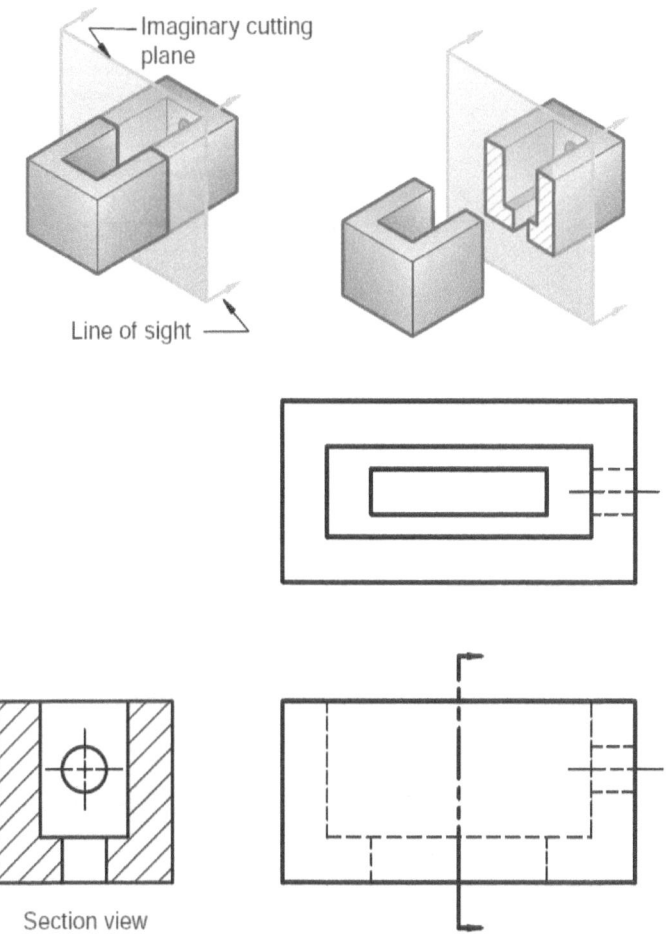

Fig. 6.6 Profile section view

Multiple sections can be done on a single object, as shown in Figure 6.7. In this example, two cutting planes are used: one horizontal and the other profile. Both cutting planes appear on edge in the front view, and are represented by cutting plane lines A–A and B–B, respectively. Each cutting plane creates a section view, and each section view is drawn as if the other cutting plane did not exist.

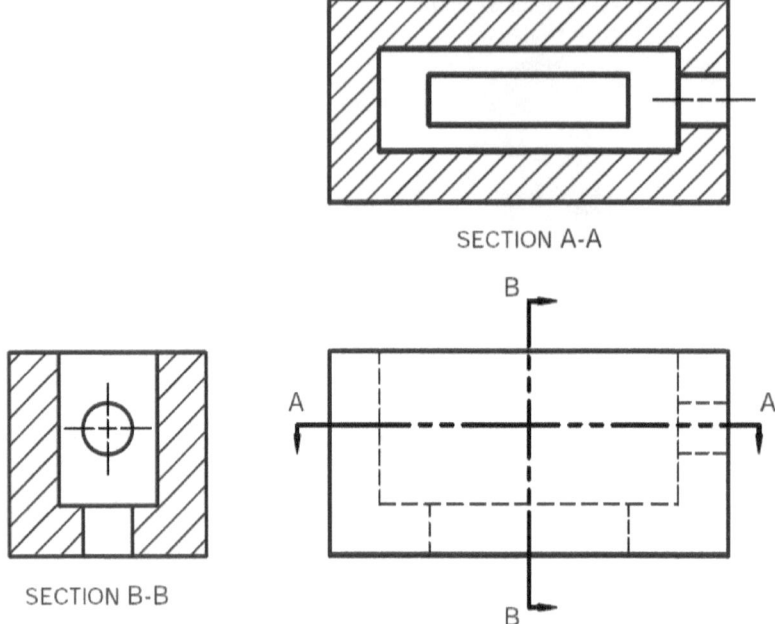

Fig. 6.6 Multiple sectioned views

6.3.2 Precedence of lines in sectional views

Cutting plane line has precedence over object line. Object line precedence over hidden line and hidden line has precedence over center line.

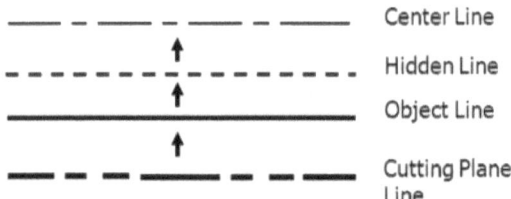

Fig. 6.7 Precedence of lines in sectional views

6.4 Section lines

Section lines or **cross-hatch lines** are added to section views to indicate the surfaces that are cut by the imaginary cutting plane. Different section line symbols can be used to represent various types of materials. However, there are so many different materials used in design that the general symbol (i.e., the one used for cast iron) may be used for most purposes on technical drawings. The actual type of material required is then noted in the title block or parts list or entered as a note on the drawing. The angle at which section lines are drawn is usually 45 degrees to the horizontal, but this can be changed for adjacent parts shown in the same section. Also, the spacing between section lines is uniform on a section view, but can be adjusted to show different parts.

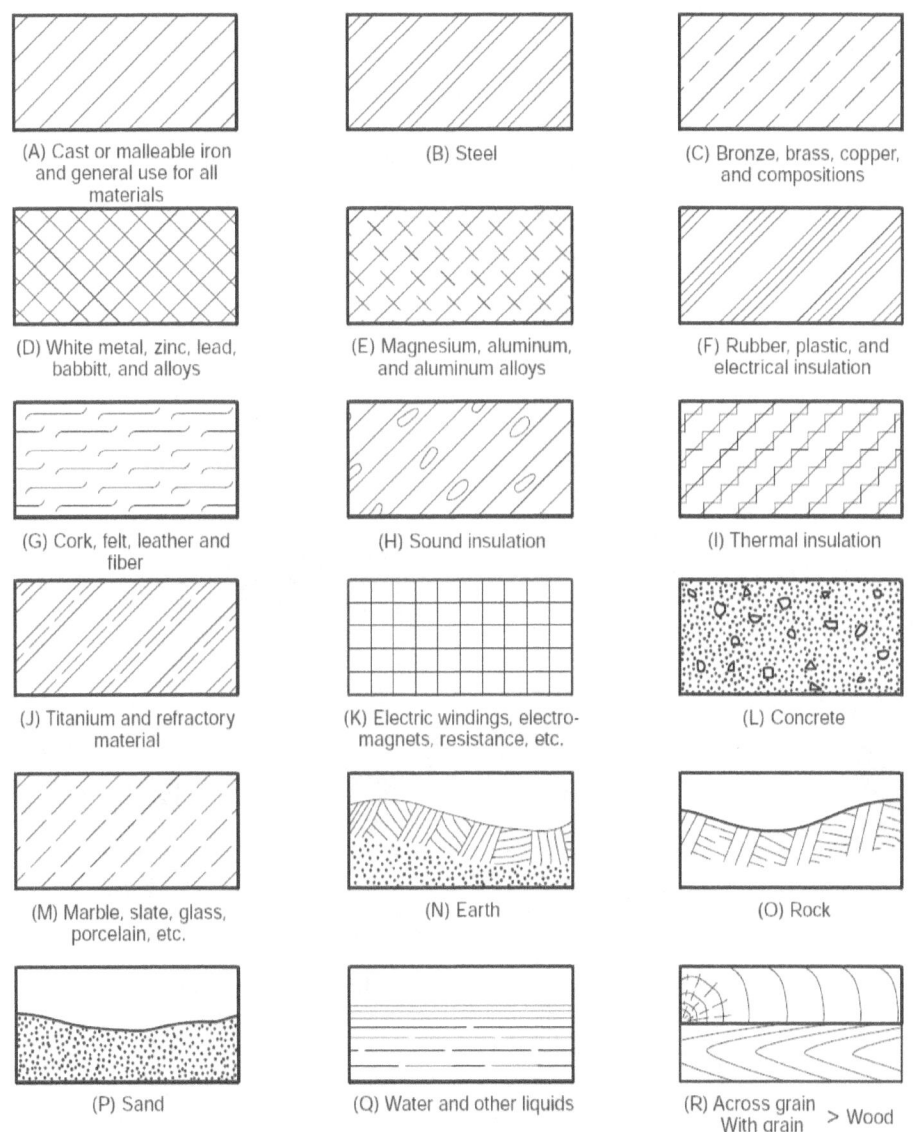

Fig. 6.8 ANSI standard section lines for various materials

The general-purpose cast iron section line is drawn at a 45-degree angle and spaced 1.5 mm to 3 mm or more, depending on the size of the drawing and part to be sectioned. As a general rule, use 3 mm spacing. Section lines are drawn as thin (.35 mm) black lines, using an H or 2H pencil.

Fig. 6.9 shows examples of good and poor section lines drawn using hand tools.

✓ The section lines should be <u>evenly spaced and of equal thickness and should be thinner than visible lines.</u>

✓ Do not run section lines beyond the visible outlines or stop them too short.

✓ Section lines should <u>not run parallel or perpendicular to the visible outline</u>.

✓ If the visible outline is drawn at 45-degree angle, the section lines are drawn at a different angle, such as 30 degrees.

✓ Avoid placing dimensions or notes within the section lined areas.

✓ If the dimension or note must be placed within the sectioned area, omit the section lines in the area of the note.

✓ Avoid placing section lines parallel or perpendicular to visible lines.

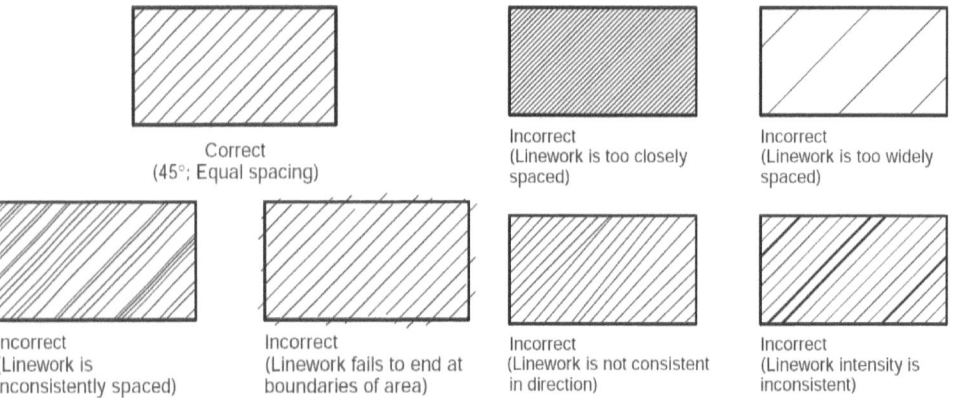

Correct
(45°; Equal spacing)

Incorrect
(Linework is too closely spaced)

Incorrect
(Linework is too widely spaced)

Incorrect
(Linework is inconsistently spaced)

Incorrect
(Linework fails to end at boundaries of area)

Incorrect
(Linework is not consistent in direction)

Incorrect
(Linework intensity is inconsistent)

Fig. 6.9 Examples of good and poor section lining techniques

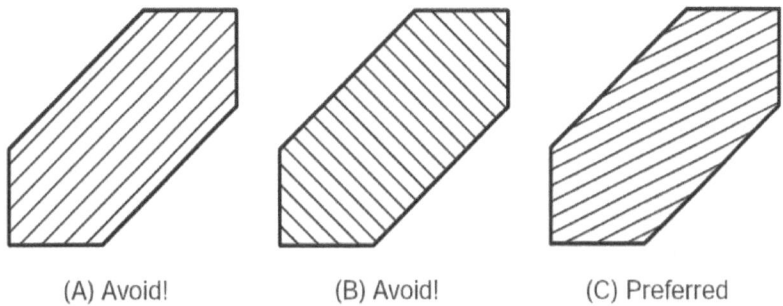

| (A) Avoid! | (B) Avoid! | (C) Preferred |

Fig. 6.10 Section lined placement. Avoid placing section lines parallel or perpendicular to visible lines.

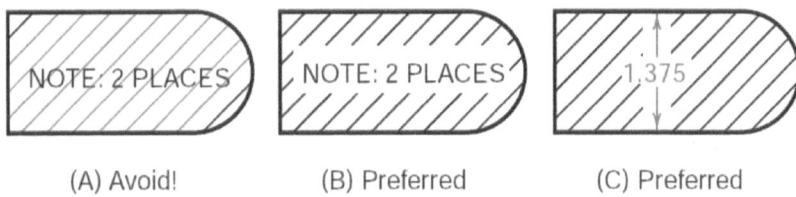

| (A) Avoid! | (B) Preferred | (C) Preferred |

Fig. 6.11 Notes in section lined areas. Section lines are omitted around notes and dimensions

Sectional views are often used to show interior features clearly for dimensioning. If dimension values or extension lines cross hatched areas, you should break the hatching behind the dimension. The best practice is to place dimensions outside the object outline.

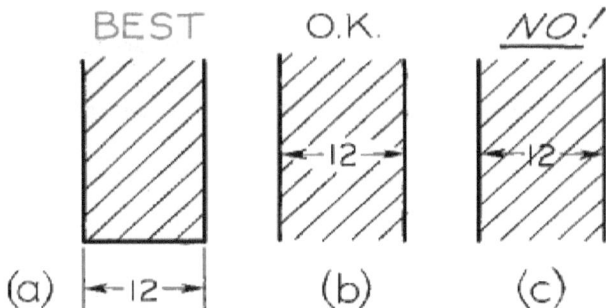

Fig. 6.12 Dimensions in sectional views

145

6.5 Types of sectional views

There are many different types of section views used on technical drawings.

- ✓ Full section view
- ✓ Half section view
- ✓ Broken-out section view
- ✓ Aligned and Revolved section views
- ✓ Removed view
- ✓ Offset section view (multiple offset views)
- ✓ Auxiliary section view

6.5.1 Full Sectional views

A full section is made by **passing the imaginary cutting plane completely through the object**. All the hidden features intersected by the cutting plane are represented by visible lines in the section view. Surfaces touched by the cutting plane have section lines drawn at a 45-degree angle to the horizontal. Hidden lines are omitted in all sectional views unless they must be used to provide a clear understanding of the object.

The top view of the sectional drawing shows the cutting plane line, with arrows pointing in the direction of the line of sight necessary to view the sectioned half of the object. In a multiview drawing, a full section view is placed in the same position that an unsectioned view would normally occupy; that is, a front section view would replace the traditional front view.

In general, the following points should be noted when making full sectional view of an object:

- ✓ In making the sectioned view, one half of the object is imagined to be removed.
- ✓ Invisible lines behind the cut surfaces are usually omitted
- ✓ Visible lines behind the section should be drawn
- ✓ Only the surfaces actually cut by the section plane are crosshatched

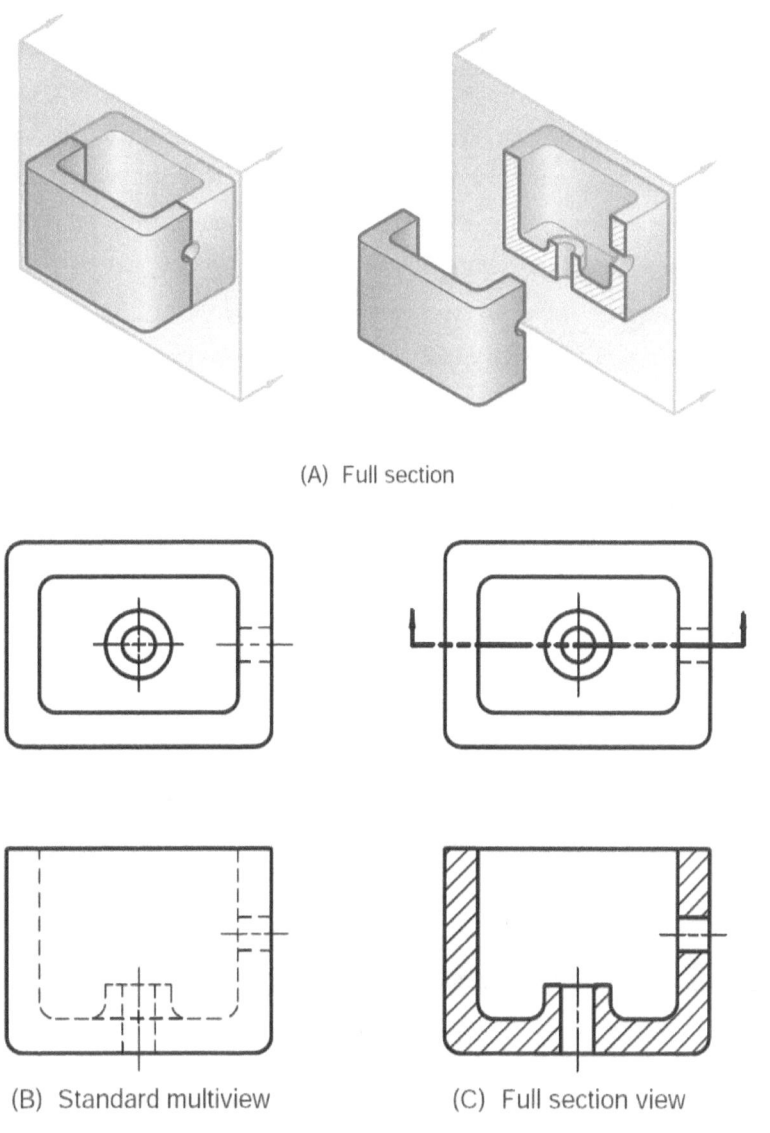

(A) Full section

(B) Standard multiview   (C) Full section view

Fig. 6.13 Full section view

The following step-by-step procedure explains and illustrates how to create a full-section view of the object shown in Fig. 6.14.

Step 1. The first step is to visualize and determine the appropriate position for the section view line from which to best view and understand the object's true internal shape and form. the imaginary cutting plane then is passed through the object through the position visualized. In this example, the cutting plane is positioned so that it passes through the centers of both holes and perpendicular to the axes of the holes. This would slice the object completely in half, revealing the interior features when viewed from the front view.

Step 2. Sketch the top view of the multiview drawing of the object.

Step 3. In the multiview, the top view of the object would be drawn as a normal view with the exception of the addition of a cutting plane line. The position of the cutting plane line would be located in the top view where the imaginary cutting plane appears on edge, which is through the horizontal center lines of the two holes in the top view. The ends of the cutting plane line are bent at 90 degrees in the direction of the line of sight necessary to create a front view with arrows added that will point away from the location of the front view.

Step 4. Visualize and sketch the front view in full section. This can be done using any of the techniques shown earlier in the chapter. For this example, we will start by sketching lines from all the edges of the object from the top view. Height dimensions in the front view are determined, then the shape of the front view is blocked in to create all the details for the front multiview drawing of the object. Lines that normally would be drawn as hidden lines will be represented as solid lines as the full-sectioned front view reveals the hidden features of the object.

Step 5. Section lines are then added to the object to those areas of the front view where the imaginary cutting plane would touch the surface or is coplanar with object surfaces. All the section lines will be spaced equally and running in the same direction because this is a single object.

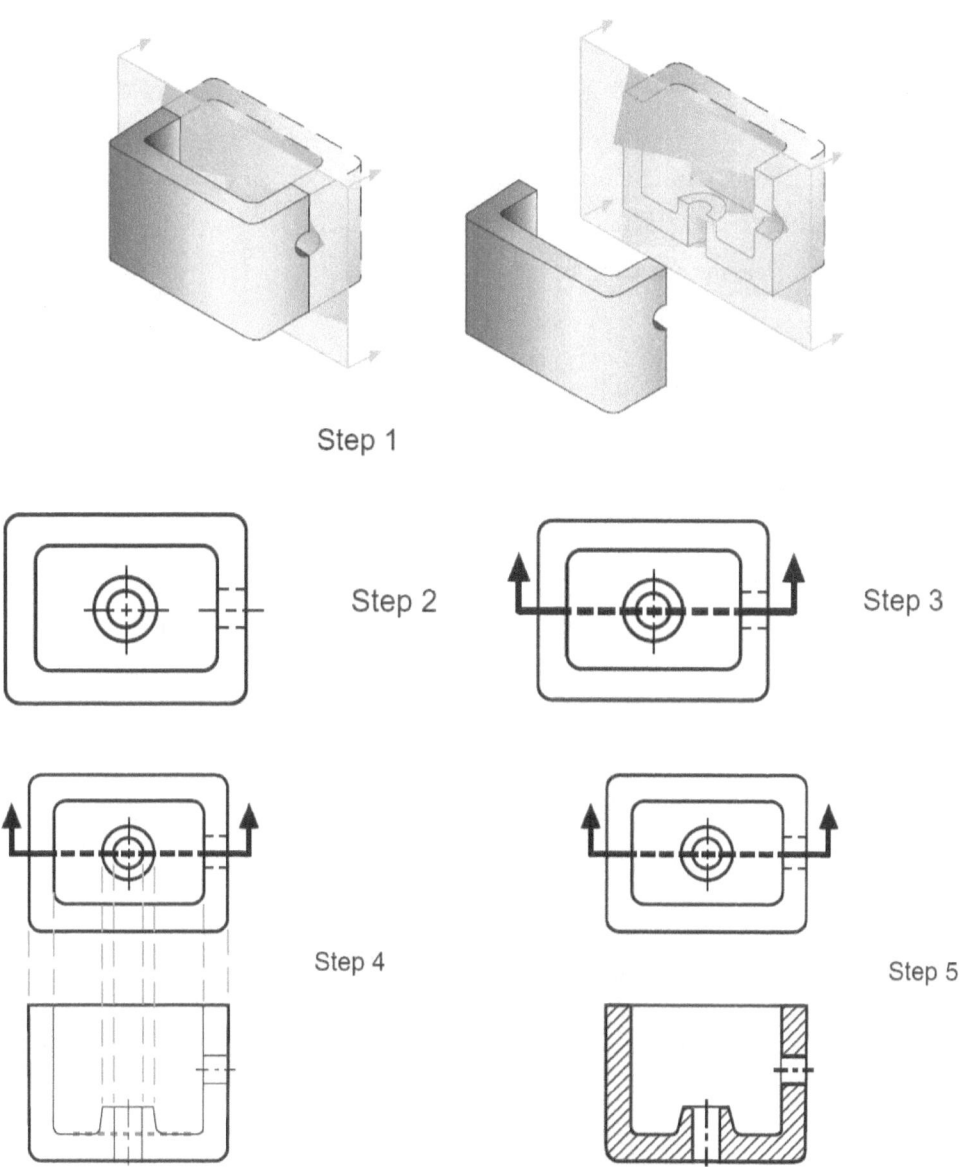

Fig. 6.14 Creating a full-section view

6.5.2 Half Sectional views

Half sectional views are created by passing an imaginary cutting plane only half way through an object. Hidden lines are omitted on both halves of the sectional view. Hidden lines may be added to the unsectioned half for dimensioning or for clarity. External features of the part are drawn on the unsectioned half of the view. A center line, not an object line, is used to separate the sectioned half from the unsectioned half of the view.

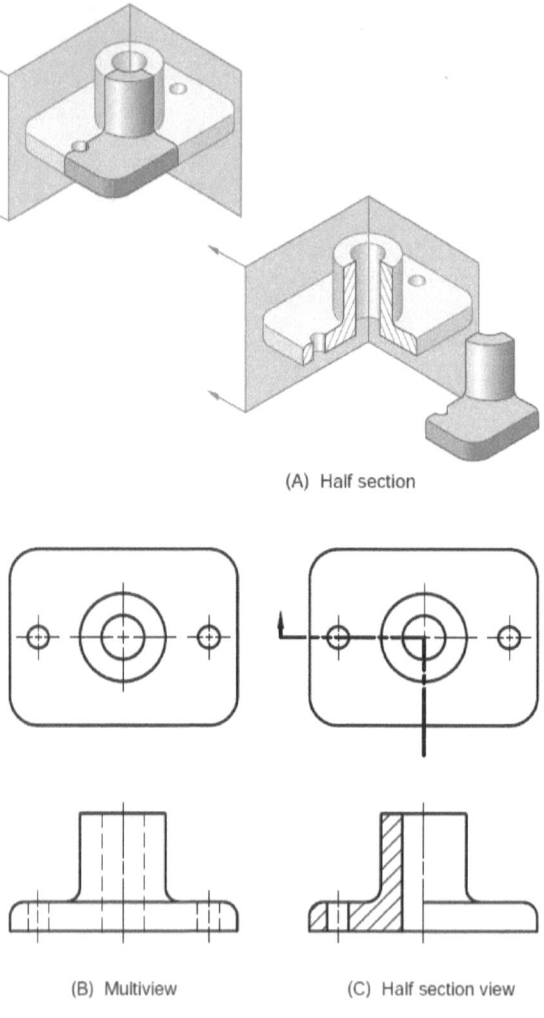

(A) Half section

(B) Multiview   (C) Half section view

Fig. 6.15 Half section

The cutting plane line shown in the top view of Fig. 6.15 is bent at 90 degrees, and one arrow is drawn to represent the line of sight needed to create the front view in section. Half section views are used most often on parts that are symmetrical, such as cylinders. Also, half sections are sometimes used in assembly drawings when external features must be shown.

6.5.3 Broken-Out Sections

A broken-out section is used when only a portion of the object needs to be sectioned. A broken out section is used instead of a half or full section to save time.

A break line separates the sectioned portion from the unsectioned portion of the view. A break line is drawn freehand to represent the jagged edge of the break. No cutting plane line is drawn. Hidden lines may be omitted from the unsectioned part of the view unless necessary for clarity.

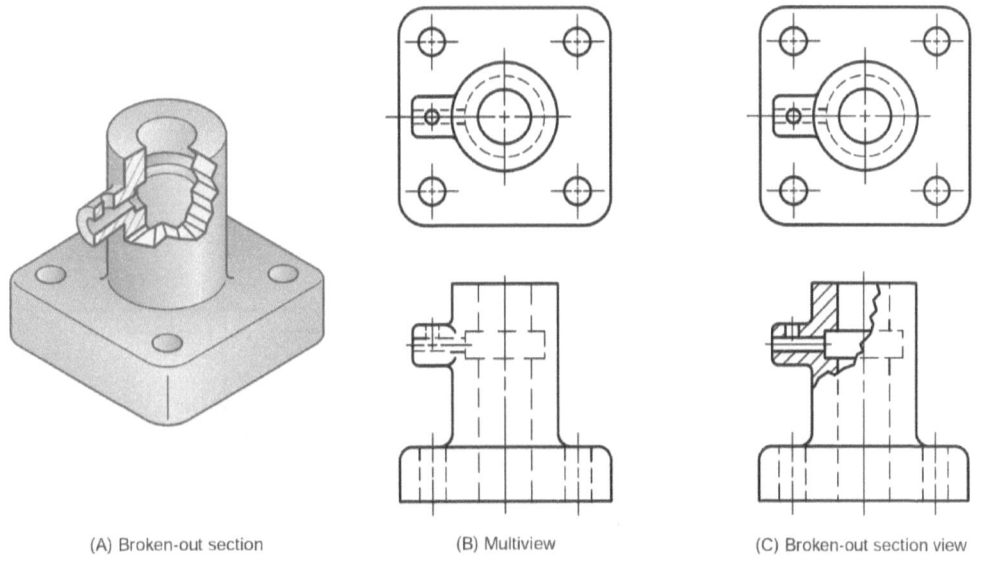

(A) Broken-out section   (B) Multiview   (C) Broken-out section view

Fig. 6.16 Broken-out section

6.5.4 Revolved Sectional views

A Revolved Sectional view is made by revolving the cross-sectional view 90 degrees about an axis of revolution and superimposing the sectional view on the orthographic view. Note that the true shape of the revolved section must be retained after the revolution regardless of the direction of the lines. Visible lines adjacent to the revolved view can either be drawn or broken out using conventional breaks.

When the revolved view is superimposed on the part, the original lines of the part behind the section are deleted. The cross section is drawn true shape and size, not distorted to fit the view. The axis of revolution is shown on the revolved view as a center line. Revolved sections are useful for describing a cross section without having to draw another view.

A revolved section is used to represent the cross section of a bar, handle, spoke, web, aircraft wing, or other elongated feature. Revolved sections are useful for describing a cross section without having to draw another view. In addition, these sections are especially helpful when a cross section varies or the shape of the part is not apparent from the given orthographic views.

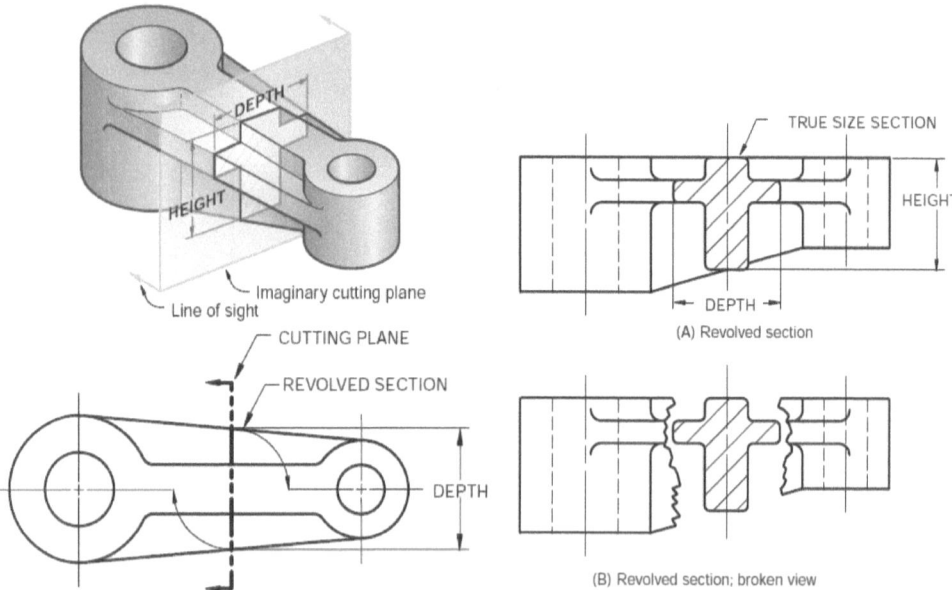

Fig. 6.17 Revolved section. A revolved section view is created by passing a cutting plane through the object, then revolving the cross section 90 degrees

152

When revolved section views are used, normally end views are not needed on a multiview drawing. A revolved section is created by drawing a center line through the shape on the plane to represent in section. Visualize the cross section of the part being rotated 90 degrees about the center line and the cross section being superimposed on the view.

If the revolved section view does not interfere or create confusion on the view, then the revolved section is drawn directly on the view using visible lines. If the revolved section crosses lines on the view it is to be revolved, then the view is broken for clarity. Section lines are added to the cross section to complete the revolved section. Visible lines adjacent to the revolved view can be either drawn or broken out using conventional breaks, as shown in Fig. 6.17B.

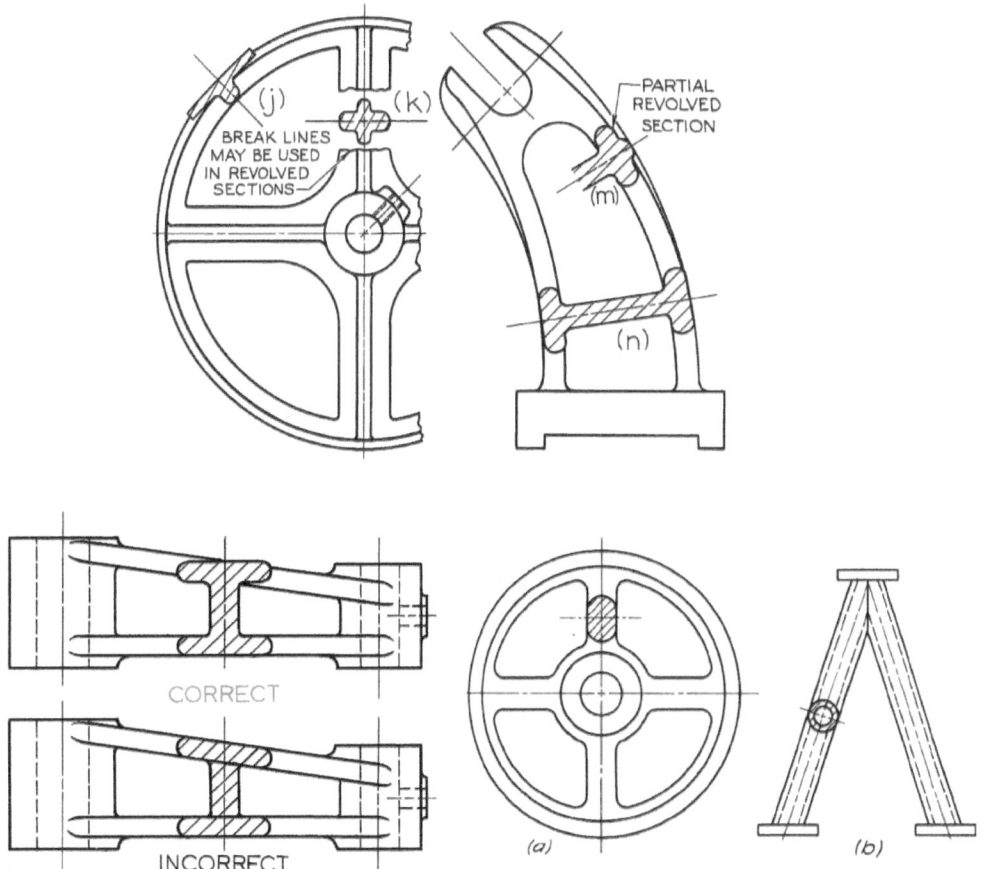

Fig. 6.18 Examples of Revolved sections

153

6.5.5 Removed Sectional views

Removed sectional views are made in a manner similar to revolved sections by passing an imaginary cutting plane perpendicular to a part then revolving the cross section 90 degrees. However, the cross section is then drawn adjacent to the orthographic view, not on it.

Removed sections are used when there is not enough room on the orthographic view for a revolved section. Removed sections are used to show the contours of complicated shapes, such as the wings of airplane, blades for jet engines or power plant turbines, and other parts that have continuously varying shapes.

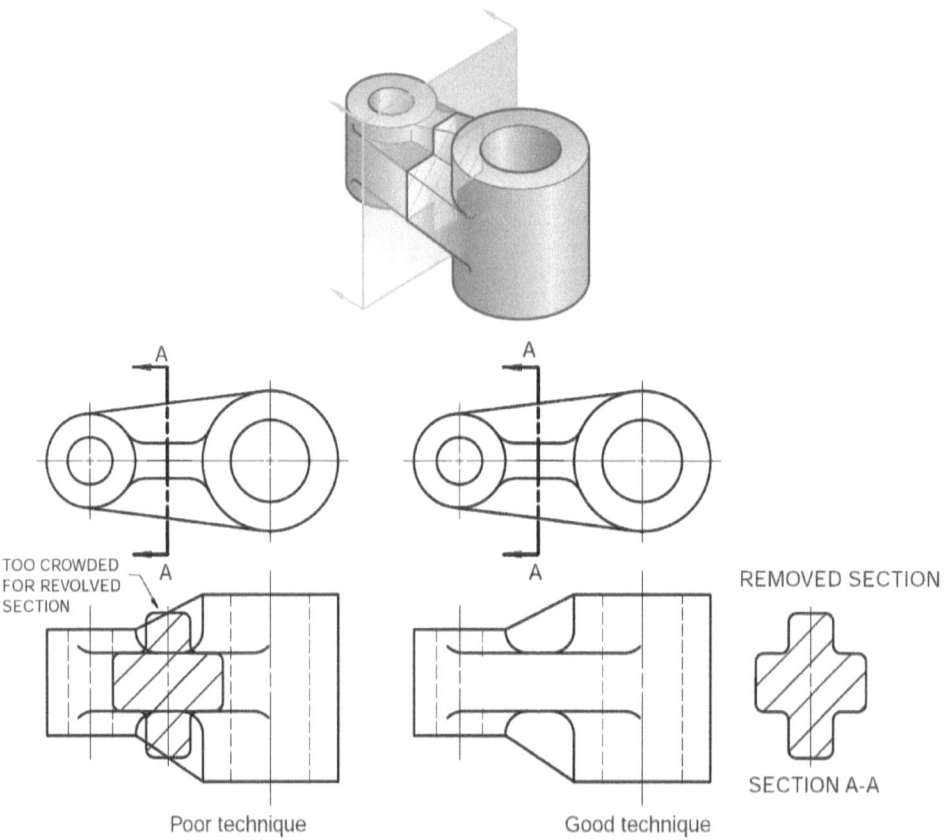

Fig. 6.19 Removed section. A removed section view is created by making a cross section, then moving it to an area adjacent to the view.

6.5.6 Offset Sectional views

An offset section has a cutting plane that is bent at one or more 90-degree angles to pass through important features. Offset sections are used for complex parts that have a number oOf important features that cannot be sectioned using a straight cutting plane.

In Fig. 6.20, the cutting plane is first bent at 90 degrees to pass through the hole and then bent at 90 degrees to pass through the slot. The front portion of the object is "removed" to create a full section view of the part. The cutting plane line is drawn with 90-degree offsets, as shown in Fig. 6.20A. As shown in Fig. 6.20B, the change of plane that occurs when the cutting plane is bent at 90 degrees is not represented with lines in the section view.

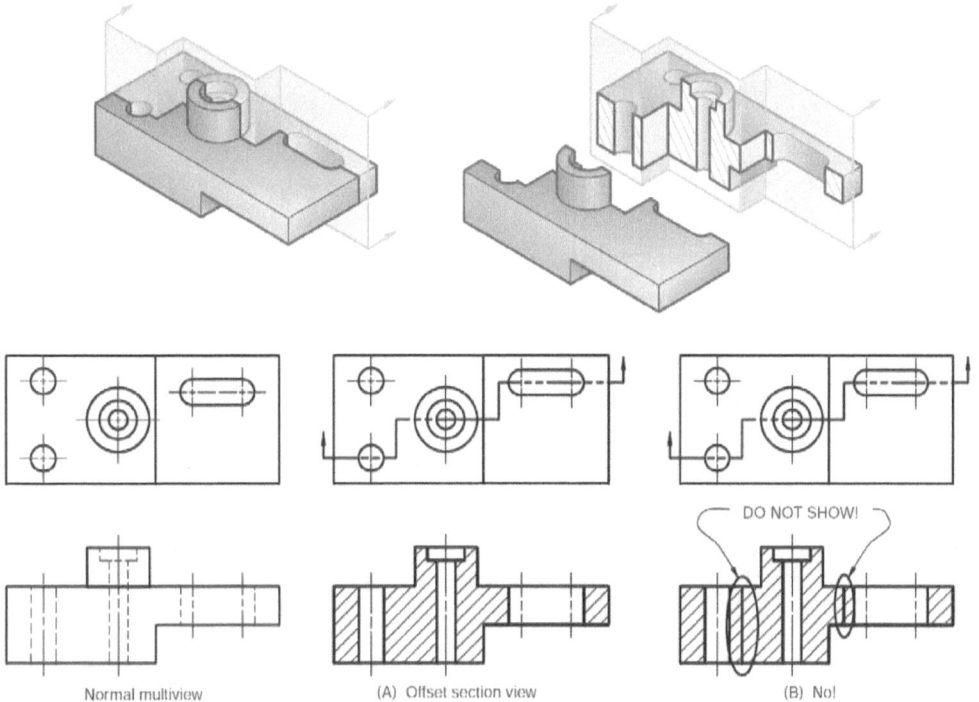

Normal multiview   (A) Offset section view   (B) No!

Fig. 6.20 Offset section. An offset section view is created by bending the cutting plane at 90-degree angles to pass through important features.

155

6.6 Special Sectioning Conventions

6.6.1 Ribs, Webs, and Other Thin Features

Ribs, webs, spokes, lugs, gear teeth, and other thin features are not section lined <u>when the cutting plane passes parallel to the feature</u>. A **rib** or **web** is a thin, flat part that acts as a support. Adding section lines to these features would give a false impression that the part is thicker than it really is. Leave the hatching of such features even though the cutting plane passes through them. Leaving thin features unsectioned only applies if the cutting plane passes parallel to the feature. <u>If the cutting plane passes perpendicular or crosswise to the feature, section lines are added</u> (Fig. 6.21 cutting plane A-A).

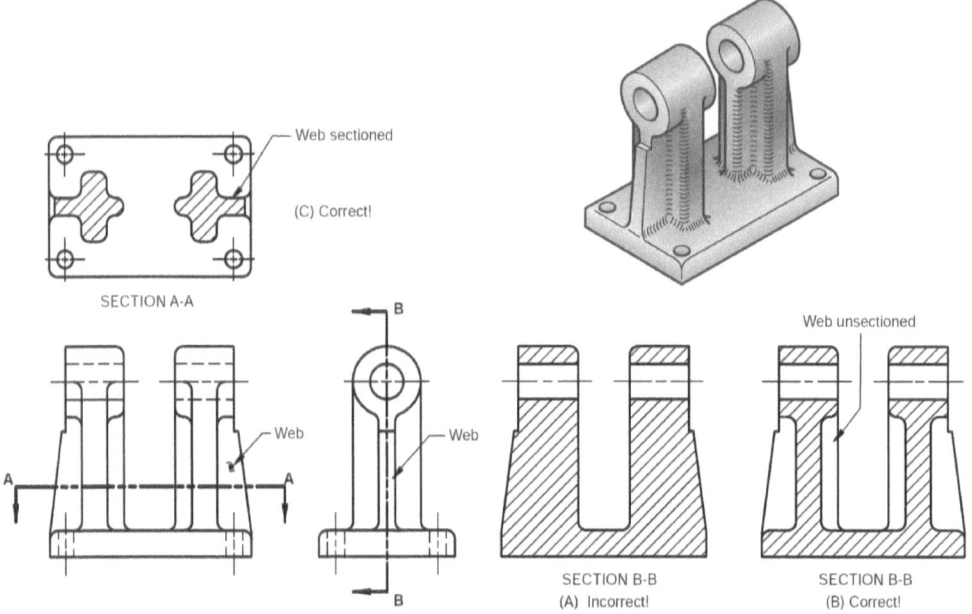

Fig. 6.21 Conventional practices for webs in section. Thin features, such as webs, are left unsectioned when cut parallel to the feature by the cutting plane

Specific features of an object are commonly left unsectioned in a sectional view if the cutting-plane line passes through and parallel to the feature. The types of features that are left unsectioned

156

for clarity are bolts, nuts, rivets, screws, rods, shafts, ribs, webs, spokes, bearings, gear teeth, pins, and keys.

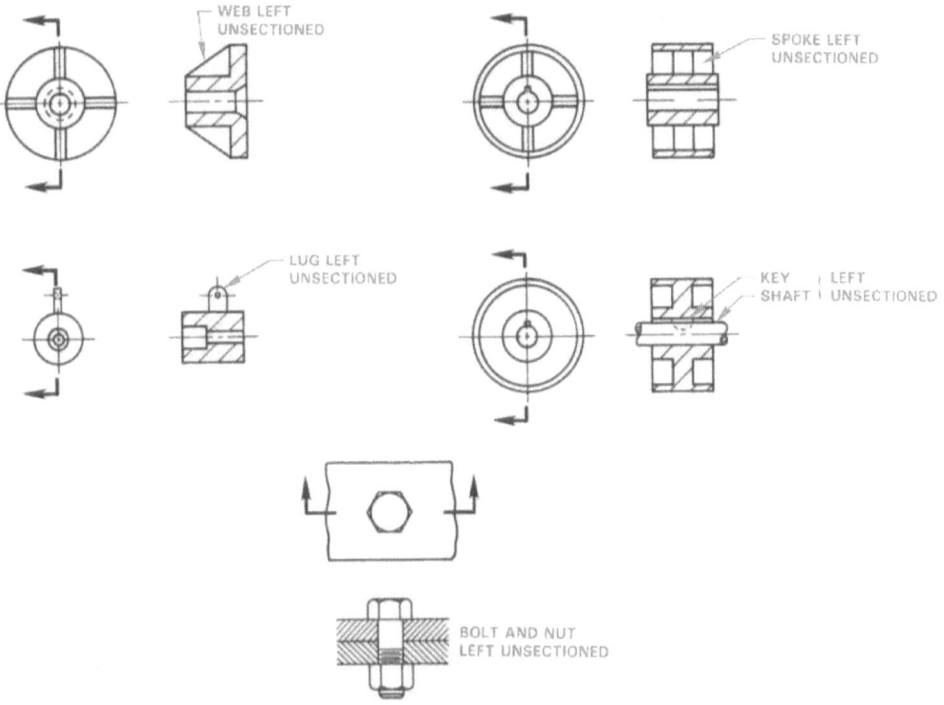

Fig. 6.22 Certain features are not sectioned when a cutting plane passes parallel to their axes. These features include webs, lugs, spokes, shafts, keys, and fasteners

6.6.2 Aligned Sections

Aligned sections are special types of orthographic drawings used to revolve or align special features of parts to clarify them or make them easier to represent in section. Aligned sections are used when it is important to include details of a part by "bending" the cutting plane. The cutting plane and the feature are imagined to be aligned or revolved before the section view is created. In other words, the principles of orthographic projection are violated in order to more clearly represent the features of the object.

Normally, the alignment is done along a horizontal or vertical center line, and the realignment is always less than 90 degrees (Fig. 6.23). The aligned section view gives a clearer, more complete description of the geometry of the part. The cutting plane line may be bent to pass through all of the nonaligned features in the unsectioned view.

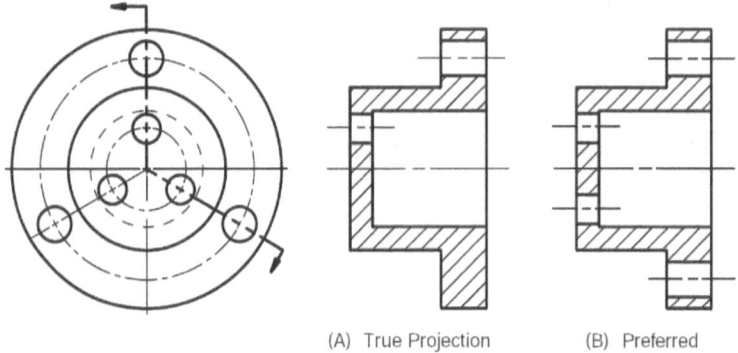

(A) True Projection   (B) Preferred

Fig. 6.23 Aligned section. Aligned section conventions are used to rotate the holes into position along the vertical center line

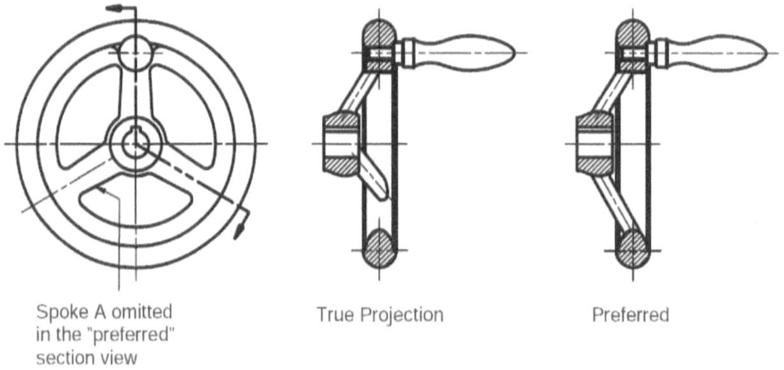

Spoke A omitted in the "preferred" section view   True Projection   Preferred

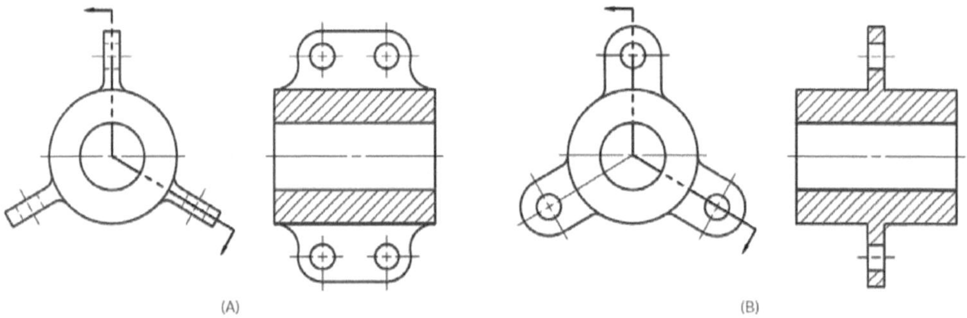

(A)   (B)

158

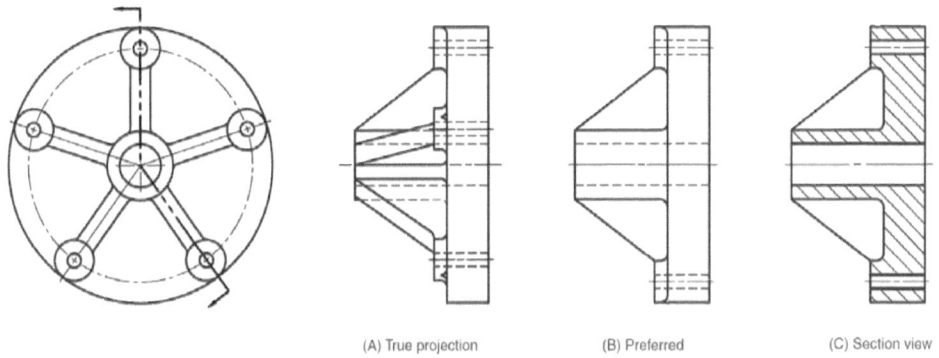

(A) True projection   (B) Preferred   (C) Section view

Fig. 6.24 Aligned section. Aligning spokes, lugs and ribs in section views is the conventional method of representation

6.6.3 Conventional Breaks

Conventional breaks are used for revolved section views or for shortening the view of an elongated part, such as a shovel handle or vehicle axle. Shortening the length of the part leaves room for the part to be drawn to large scale. Cylindrical or tubular material is represented as a figure eight or S. For very small parts, the figure eight can be drawn freehand or by using an irregular curve. Breaks for rectangular metal or wood are drawn freehand.

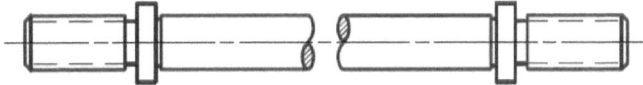

Fig. 6.25 Conventional break symbols can shorten the drawn length of a long object

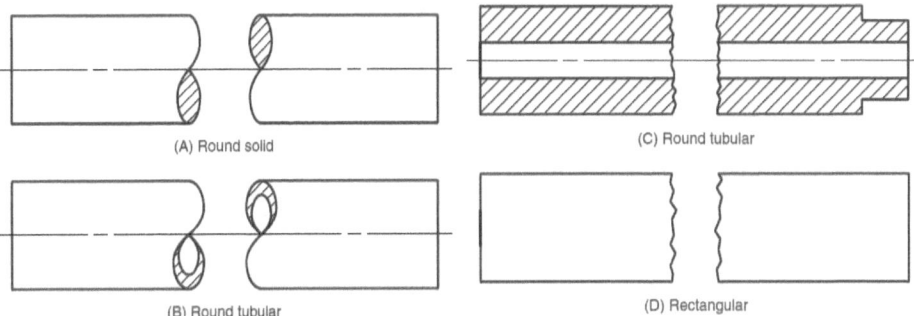

(A) Round solid

(B) Round tubular

(C) Round tubular

(D) Rectangular

Fig.6.26 Examples of conventional break symbols used for various materials

159

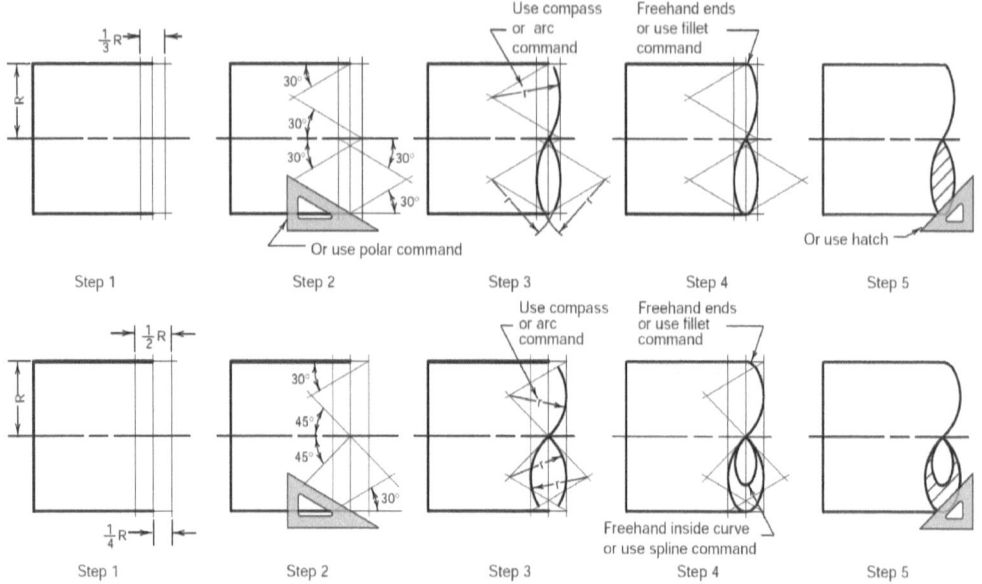

Fig. 6.27 Procedures used to create a break for cylindrical or tubular material

6.7 Problems

1. Sketch full-section views of the objects shown below. Consider each grid to be 6 mm.

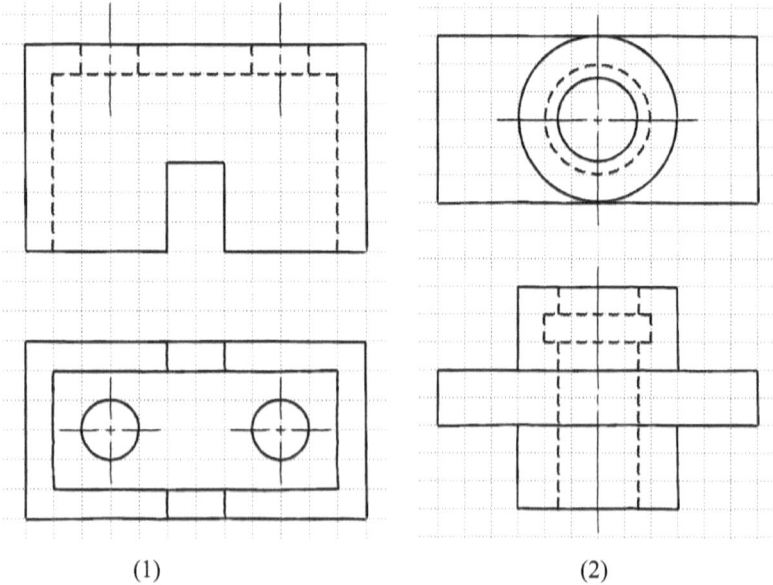

(1)   (2)

160

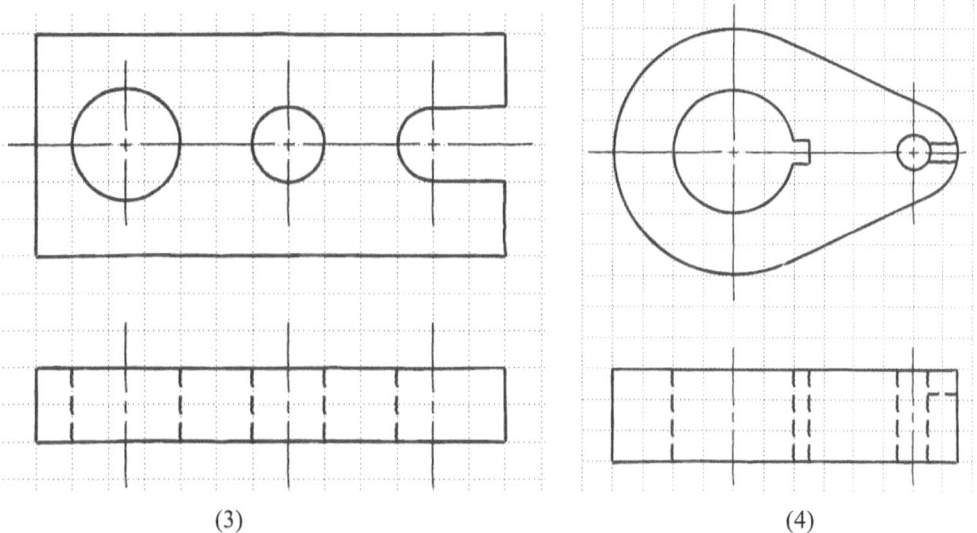

(3)   (4)

2. Sketch offset-section views of the objects shown below. Consider each grid to be 6 mm

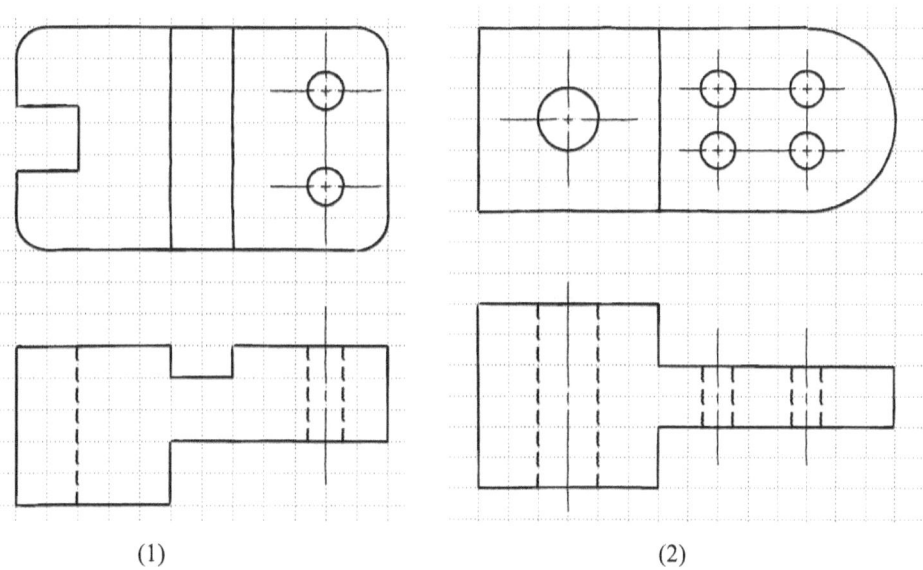

(1)   (2)

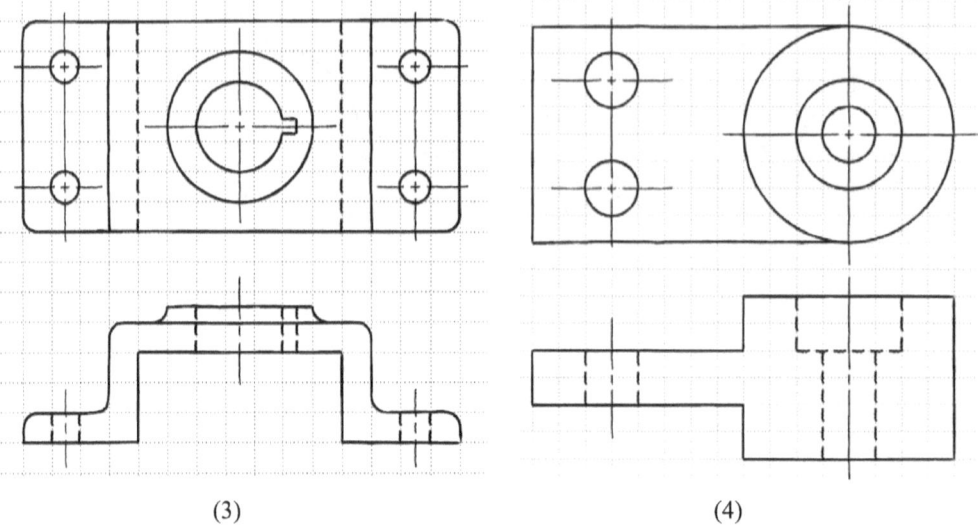

(3)   (4)

3. Sketch half-section views of the objects shown below. Consider each grid to be 6 mm

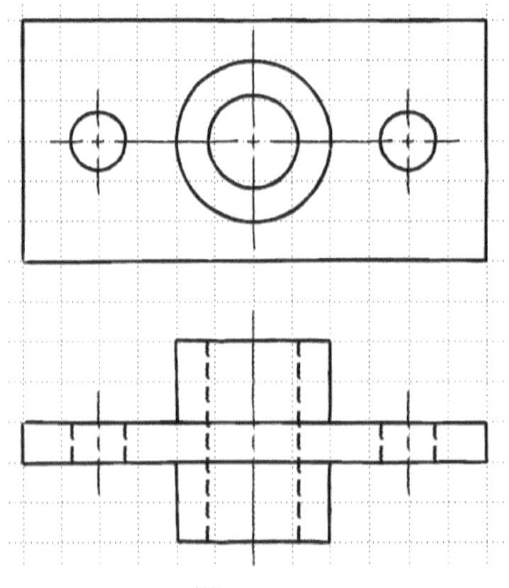

(1) 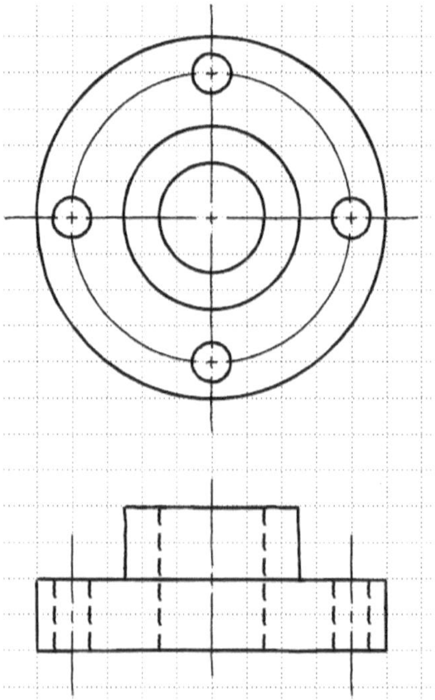   (2)

(3)

(4)

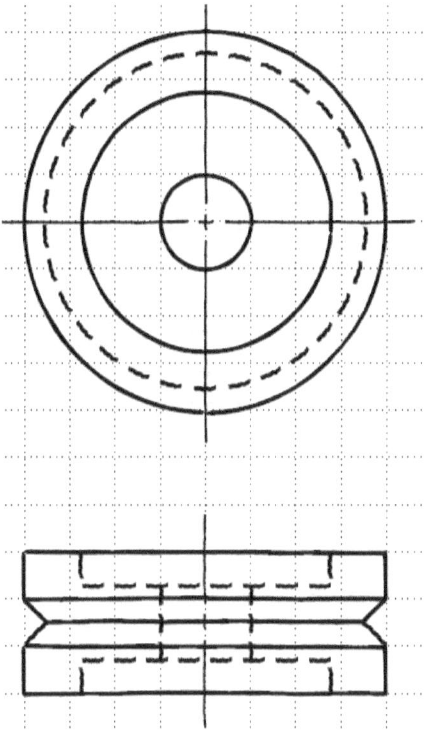

4. Sketch broken-out views of the objects shown below. Consider each grid to be 6 mm

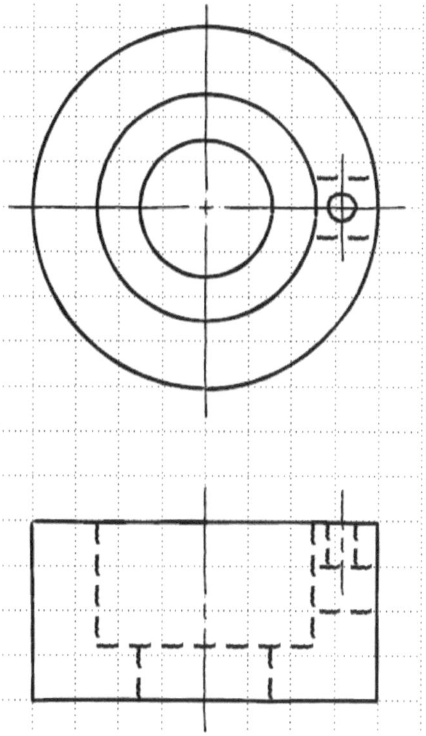

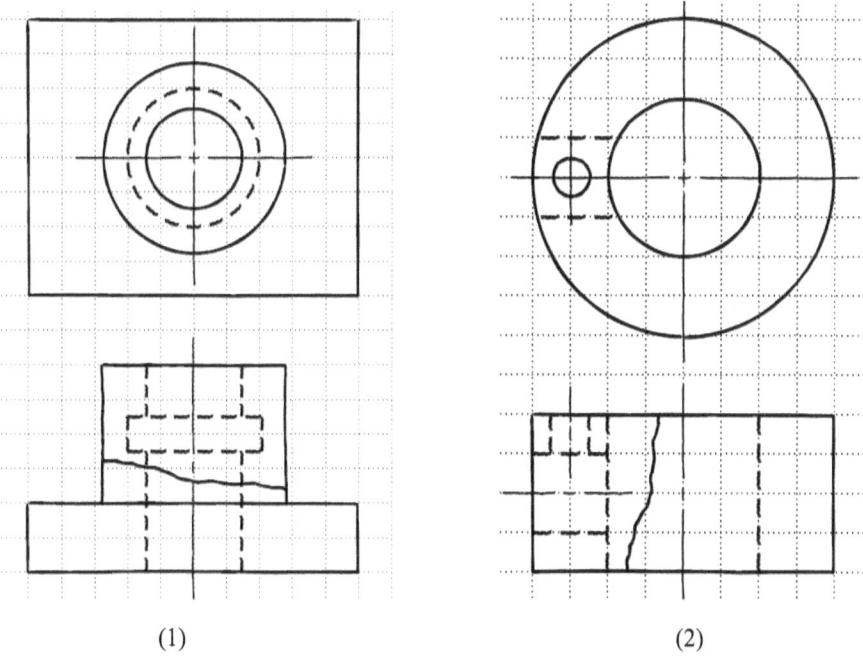

(1)

(2)

5. For the principal views shown below draw an appropriate full section of the front views

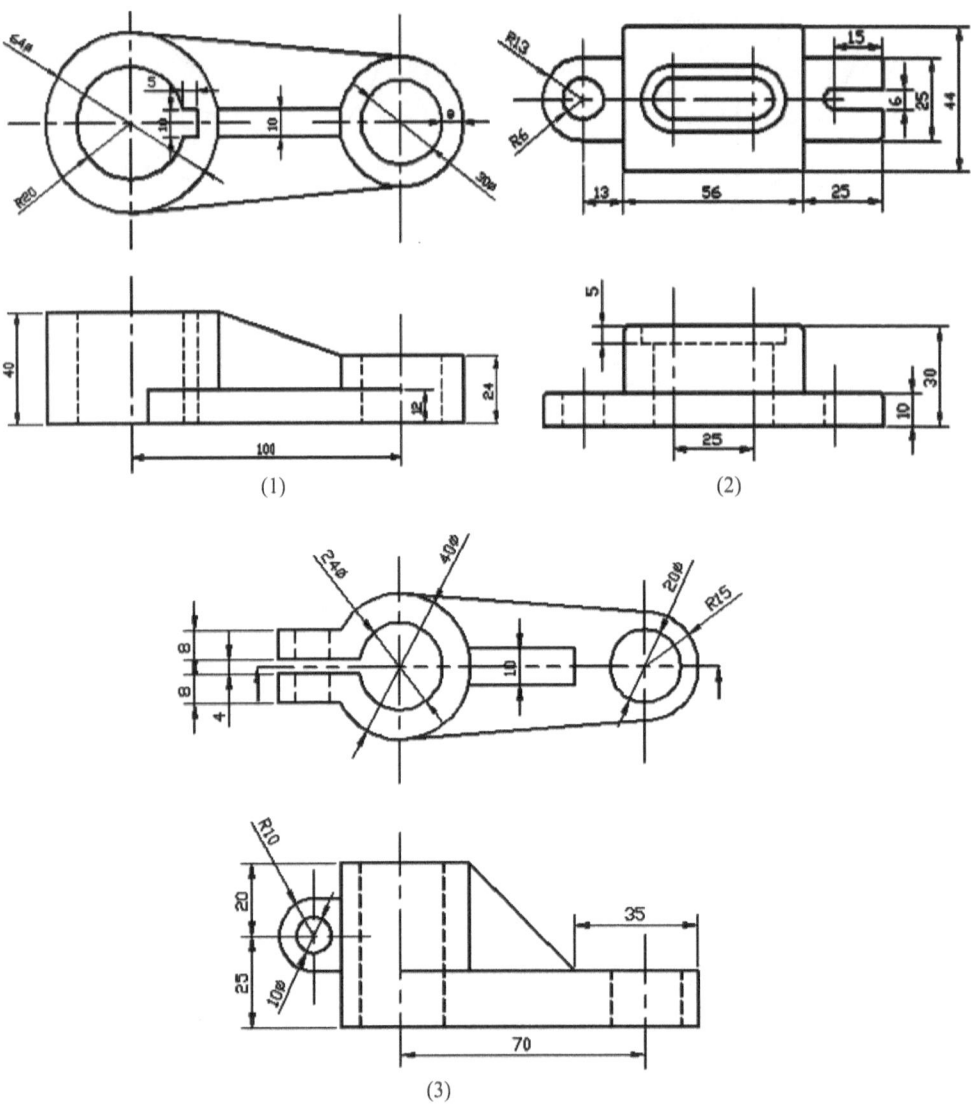

(1)

(2)

(3)

164

6. For the views shown below draw the front view as an appropriate half section

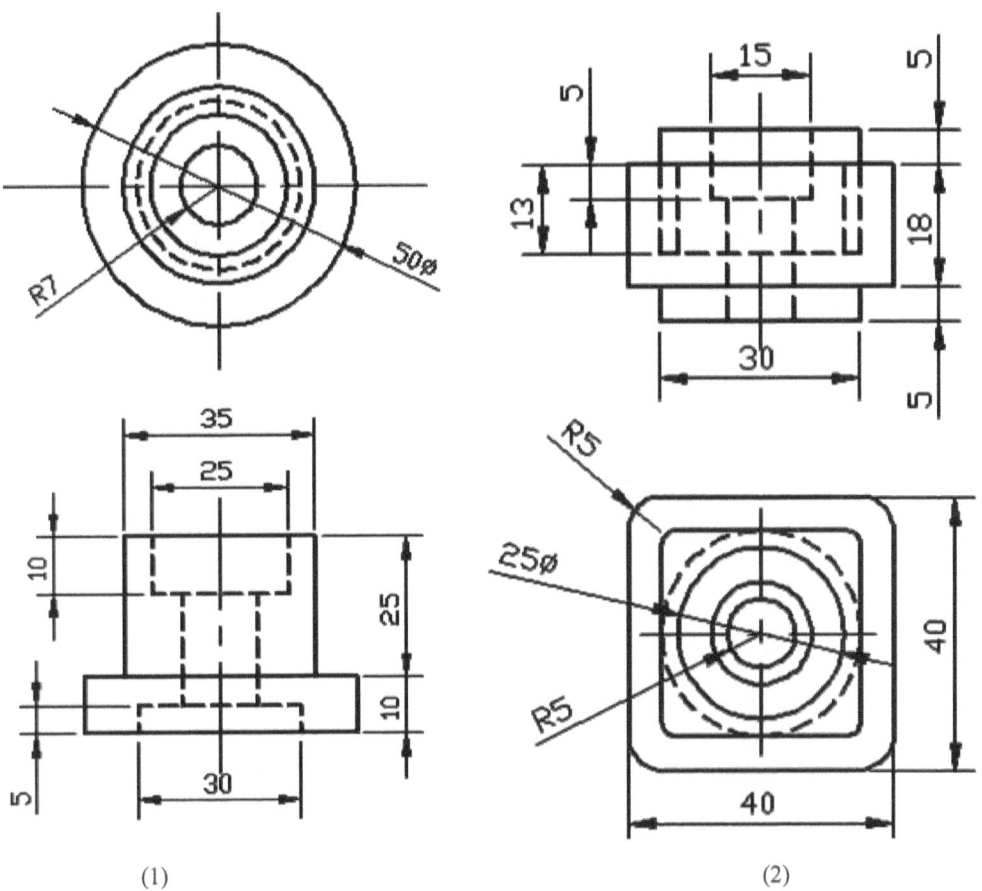

(1)   (2)

7. For the views shown below choose **appropriate offset cutting plane** line to show the front
   view in section.

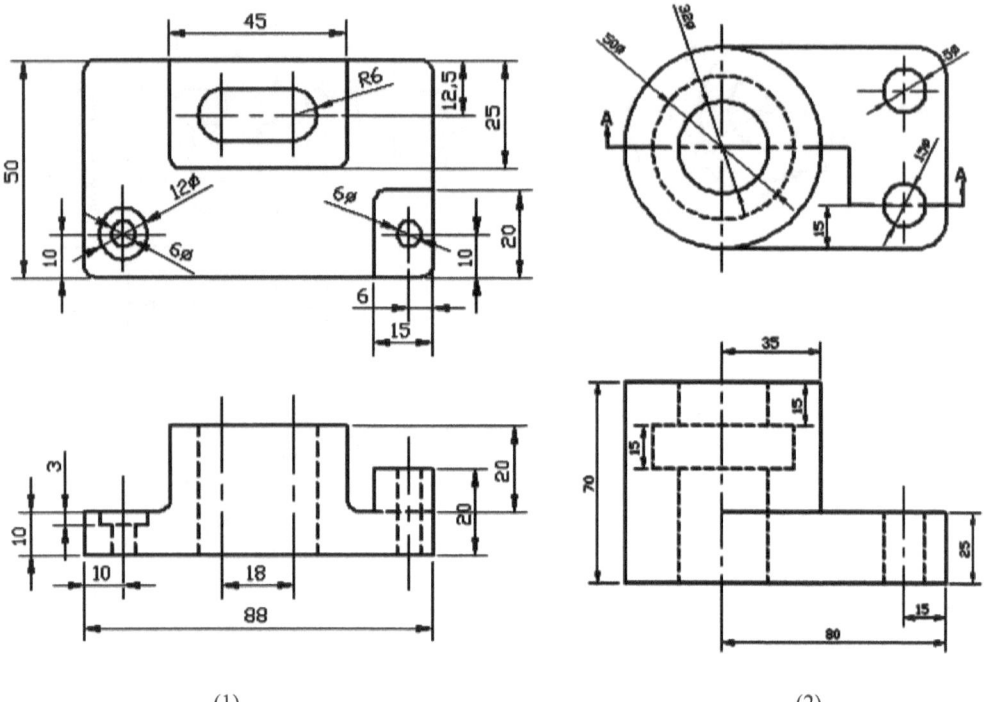

(1)   (2)

CHAPTER 7

DEVELOPMENT AND INTERSECTION OF SURFACES

7.1 Development of Surfaces

A development is the unfolded or unrolled flat or plane figure of a 3-D object. When the complete surface of a solid is opened out and laid on a plane, the surface of the solid is said to have been "developed" and the figure obtained is called **"development" of the surface.** Every line on the development must be true length of the corresponding line on the surface.

A **developable surface** is one which may be unfolded or unrolled so as to coincide with a plane. **Single curved surfaces** such as of cones and cylinders can be accurately developed. **Warped and double curved surfaces** such as of sphere, torus, ellipsoid and hyperboloid are **undevelopable**. These can, however, be approximately developed by dividing them into a number of parts.

In engineering practice an engineer is frequently required to have knowledge of development of surfaces of an object in its design and manufacturing processes. Practical applications of development occur in sheet-metal work, stone cutting and pattern making. Fig. 7.1 below shows the development of surfaces of common geometrical solids. Here the complete surface of each solid is opened out.

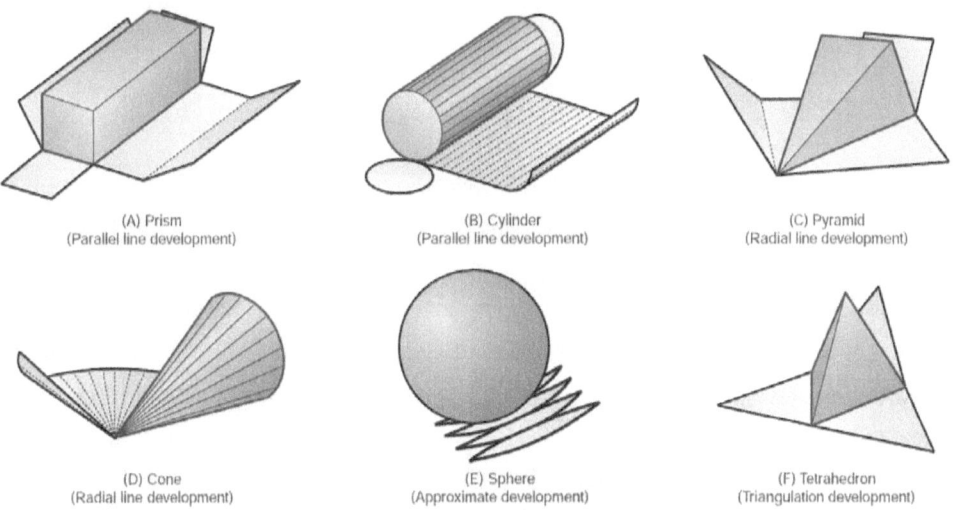

(A) Prism
(Parallel line development)

(B) Cylinder
(Parallel line development)

(C) Pyramid
(Radial line development)

(D) Cone
(Radial line development)

(E) Sphere
(Approximate development)

(F) Tetrahedron
(Triangulation development)

Fig. 7.1 Development of surfaces of geometrical solids

167

7.1.1 Principal Methods of Development

There are four principal methods of developments, based either on the type of surface or the method used to create the development.

i. Parallel-line development

Parallel-line developments are made from common solids that are composed of parallel lateral edges or elements. **Prisms and cylinders** are solids that can be flatted or unrolled into a flat pattern, and all parallel lateral surfaces or elements will retain their parallelism. It is employed in cases of prisms and cylinders in which **stretch-out line** principle is used. A stretch-out-line is the development of the end of the prism or the cylinder.

ii. Radial line development

It is employed in case of **pyramids and cones** in which the true length of the slant edge or generator is used as radius. In the development, all the elements of the solid become radial lines that have the vertex as their origin.

iii. Triangular development

It is employed where the connecting surface is neither prismatic nor pyramidal. In this method the various plane surfaces are divided into several triangular areas, and each triangle is laid out in the development after the termination of its true form.

iv. Approximate Development

It is employed where the surface of a solid is theoretically undevelopable.

7.1.2 Parallel-line Development

1. Development of Prisms

Prisms are developed based on the parallel line development method. To develop a prism, a stretch-out line, whose length is equal to the perimeter of the prism, is laid out first. Then, points are marked on this line at an interval equal to the lengths of the faces of the prism. Then, lines representing the fold lines are drawn perpendicular to the stretch-out-line through these points. Finally, the lengths of the edges are transferred from the front view of the prism.

To develop a full prism

1. **Draw** the front and top views of the prism.
2. **On** the top view, number the edges in the clockwise direction so as to ensure the development will be made inside up. Also, number the edges on the front view in agreement with the numbering on the top view.
3. **Construct** the stretch-out line 1-1 through the base of the front view.
4. **Transfer** the widths of the faces from the top view to the stretch-out-line, locating points 1, 2, 3, etc. on the stretch-out line.
5. **Draw** lines perpendicular to the stretch-out line through points 1, 2, 3, etc.
6. **Transfer** the true lengths of the edges of the prism from the front view to the corresponding line on the development, locating points 1', 2', 3', etc. To do so you may use either projectors or divider.
7. **Complete** the development by joining points 1', 2', 3', etc. Also, attach the lower base and the upper base of the prism to get the development of the entire surface of the prism.

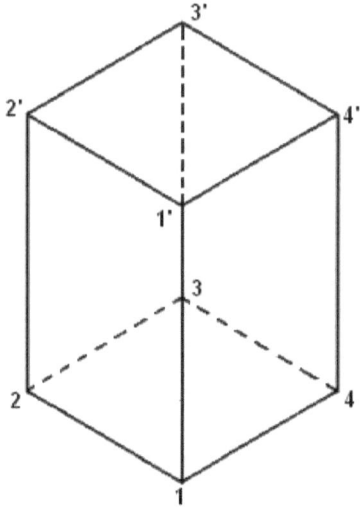

(a) A rectangular prism

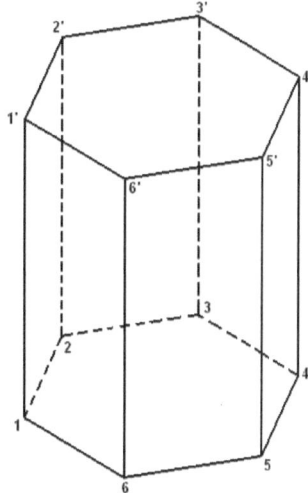

(b) Hexagonal Prism

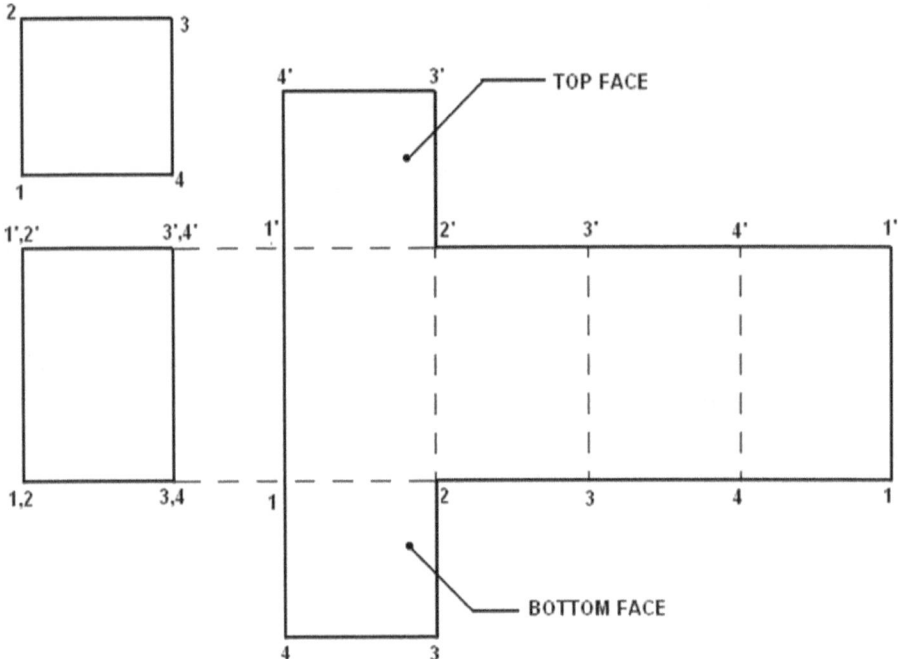

Fig. 7.2 Development of Rectangular Prism

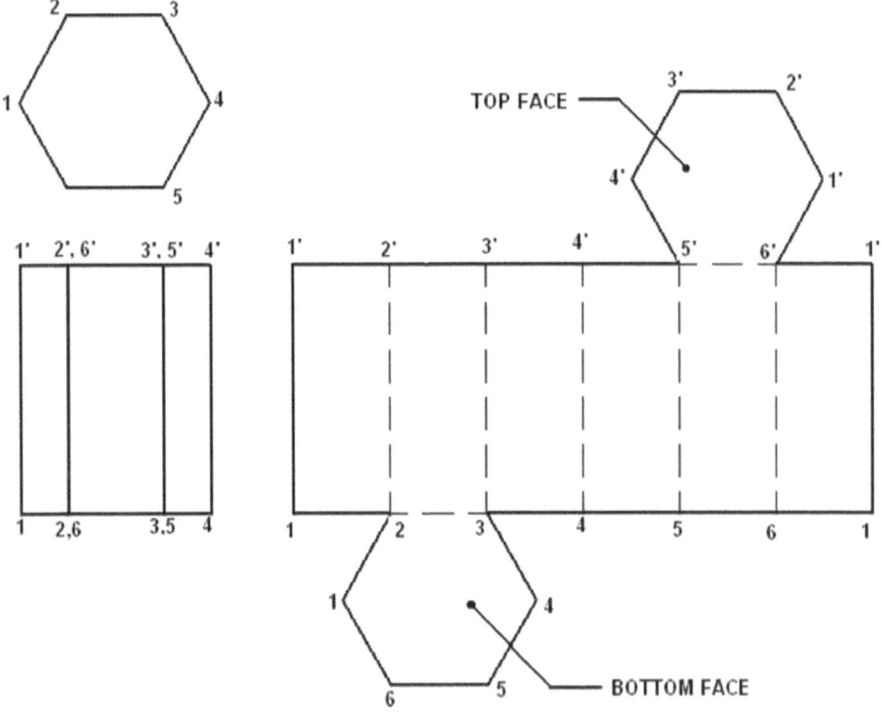

Fig. 7.3 Development of Hexagonal Prism

2. Development a truncated prism

The same basic steps are used to create a truncated prism development as were described in the full prism.

1. Draw the front and top views of the prism. Also, draw the auxiliary view of the inclined surface.
2. On the top view, number the edges in the clockwise direction, starting from 1 at the shortest edge of the prism. Also, number the edges on the front view in agreement with the numbering on the top view.
3. Construct the stretch-out line 1-1 through the base of the front view.
4. Transfer the widths of the faces from the top view to the stretch-out-line, locating points 1, 2, 3, etc. on the stretch-out line.
5. Draw lines perpendicular to the stretch-out line through points 1, 2, 3, etc.
6. Transfer the true lengths of the edges of the prism from the front view to the corresponding line drawn in step 5; locating points 1', 2', 3', etc. To do so you may use either projectors or divider.
7. Complete the development by joining points 1', 2', 3', etc. Also, attach the lower base and the auxiliary view of the inclined face of the prism to get the development of the entire surface of the prism.

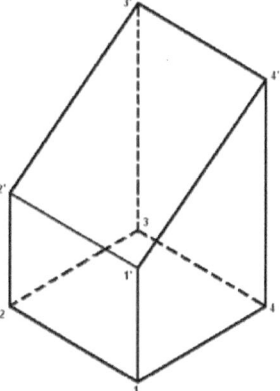

(a) A Truncated Prism

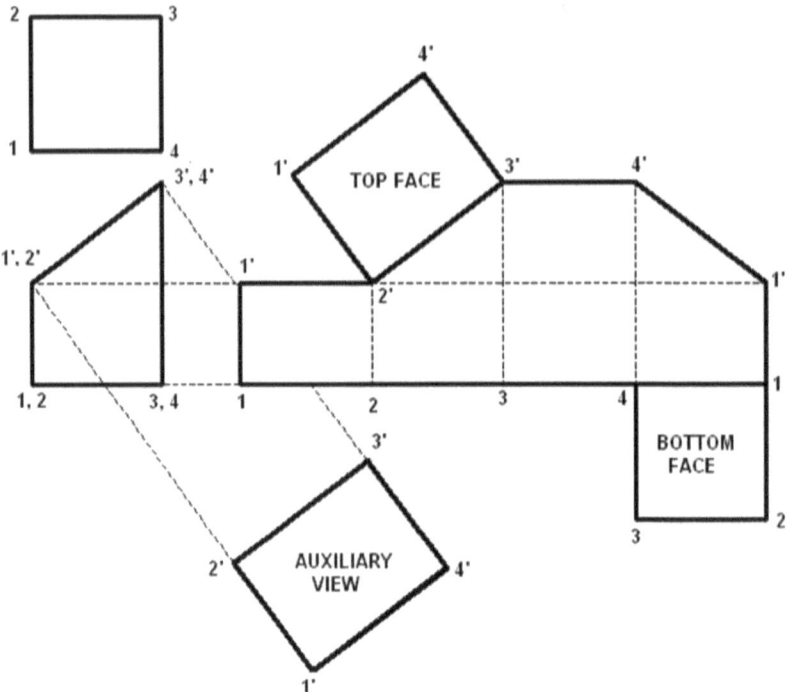

Fig. 7.4 Development of a Truncated Prism

3. Development of Cylinders

(a) To develop a Full Right Cylinder

The development of cylinders is also based on the parallel-line development method or stretch-out line method. In this case, the length of the stretch-out line is equal to the circumference of the cylinder.

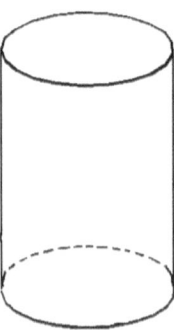

(a) Full Right Cylinder

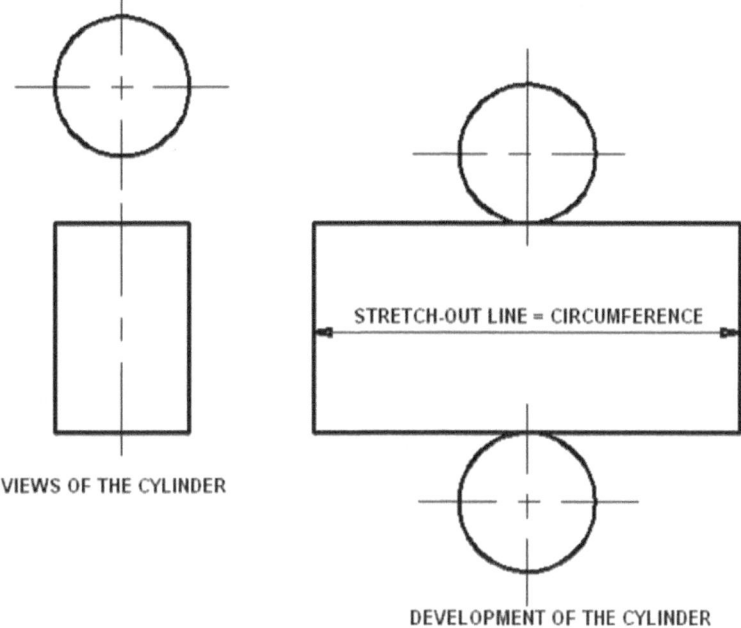

VIEWS OF THE CYLINDER

STRETCH-OUT LINE = CIRCUMFERENCE

DEVELOPMENT OF THE CYLINDER

Fig. 7.5 Development of Full Right Cylinder

(b) To develop a truncated Cylinder

1. Draw the front and top views of the truncated cylinder. Also, draw the auxiliary view of the inclined face.

2. Divide the top view of the cylinder into appropriate number of equal parts. Then, number the resulting surface line elements starting from 1 at the shortest surface element. Also, number the elements on the front view in agreement with the numbering on the top view.

3. Draw the stretch-out line and mark points A and B such that AB= $2\pi R$ (the circumference of the cylinder).

4. Divide AB into the same number of equal parts as step 2, and mark points 1, 2, 3, etc.

5. Draw perpendicular lines to the stretch-out line through points 1, 2, 3, etc.

6. Transfer the true lengths of all line elements from the front view to the corresponding line on the development to locate points 1', 2', 3', etc.

7. Draw a smooth curve through points 1', 2', 3', etc., using a French curve.

8. Attach the lower base and the inclined face to get the development of the entire

surface of the cylinder.

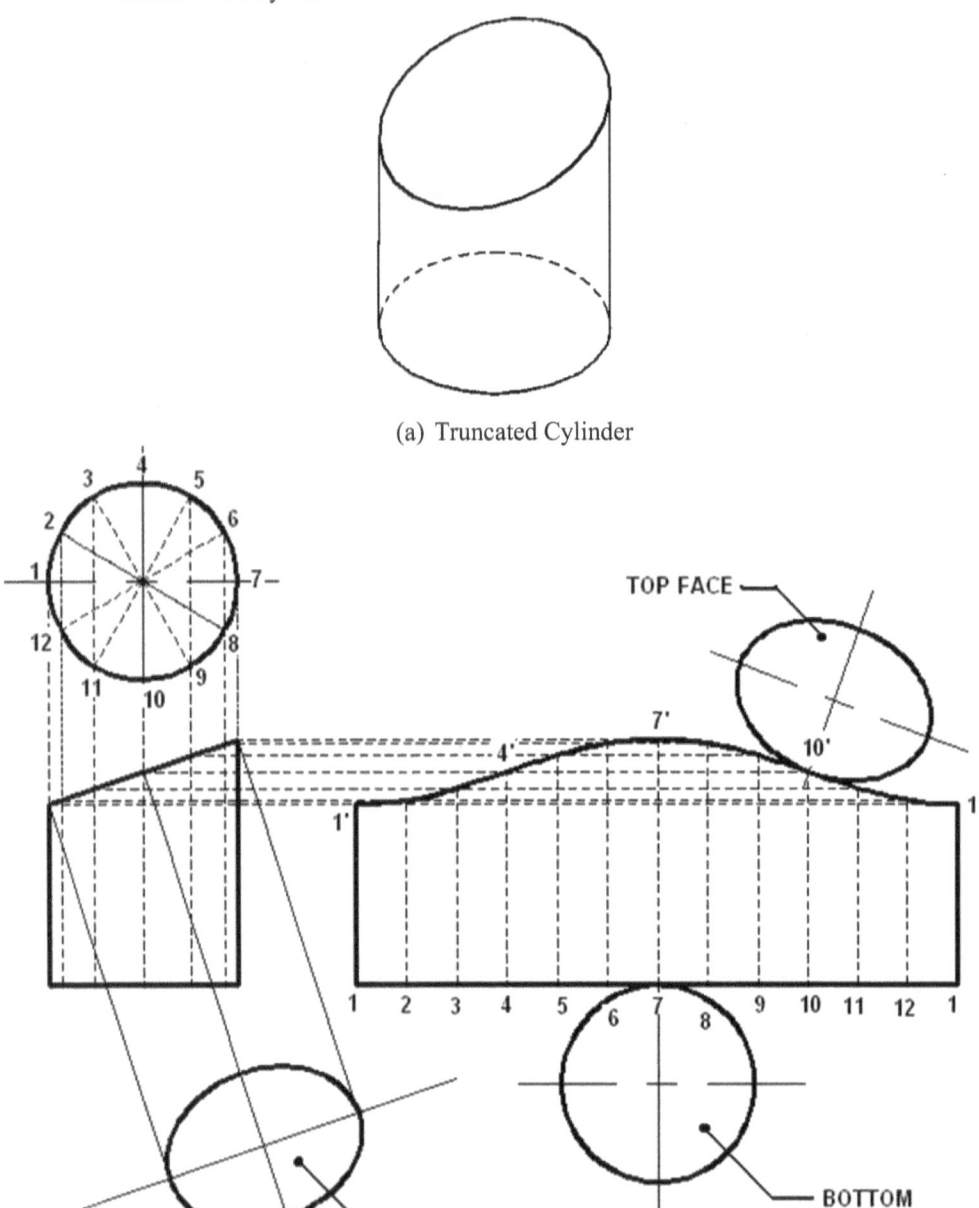

(a) Truncated Cylinder

Fig. 7.6 Development of Truncated Cylinder

7.1.3 Radial Line Development

The development of an object tapering to an apex may be made by the radial line development method. Examples of objects whose developments can be made by radial line development method are **pyramids and cones.**

The radial line development method is based on the location of a series of lines which radiate from the apex down the surface of the object to a base.

(a) Development of Pyramids

The development of a pyramid is based on the radial line development method. In the process of developing a right pyramid, a large arc is made with radius equal to the length of the edge of the pyramid. Then, points are marked on this arc using compass by setting off arcs with radius equal to the true lengths of the sides of the base of the pyramid. The true lengths of the edges of a pyramid are found by using the triangulation method.

1. Full Pyramid Development

i.   Draw the front and top views of the pyramid

ii. Number the edges of the pyramid on the top view in the clockwise direction. Also, number the edges on the front view in agreement with the numbering on the top view.

iii. Find the true lengths of the edges of the pyramid by triangulation method.

iv. Use the true lengths of one of the edges of the pyramid to draw an arc.

v.   Use the true lengths of the sides of the base as radii to draw arcs that will intersect the arc drawn in step 4 at points 1, 2, 3, etc. as shown in figure below.

vi. Join points 1, 2, 3, etc. to each other and to the center O of the arc drawn in step 4. Also, attach the base of the pyramid to complete the development.

True length by triangulation

The procedure to determine the true length of a line is the most important concept to carry out radial line development. **Triangulation method** is used to find true length of lines that are not parallel to any principal planes. Consider a line AB whose front and top views are shown below. The true length of the line is the hypotenuse of the right-angled triangle whose legs are equal to the top view of the line (a_2b_2) and the vertical height (H) of the front view. Therefore, the true length of line AB can be found by rotating the top view a_2b_2 about a_2 until it becomes horizontal. Then, b_2' is projected to the front view to locate b_1' in order to get a_1b_1' which is the true length of line AB.

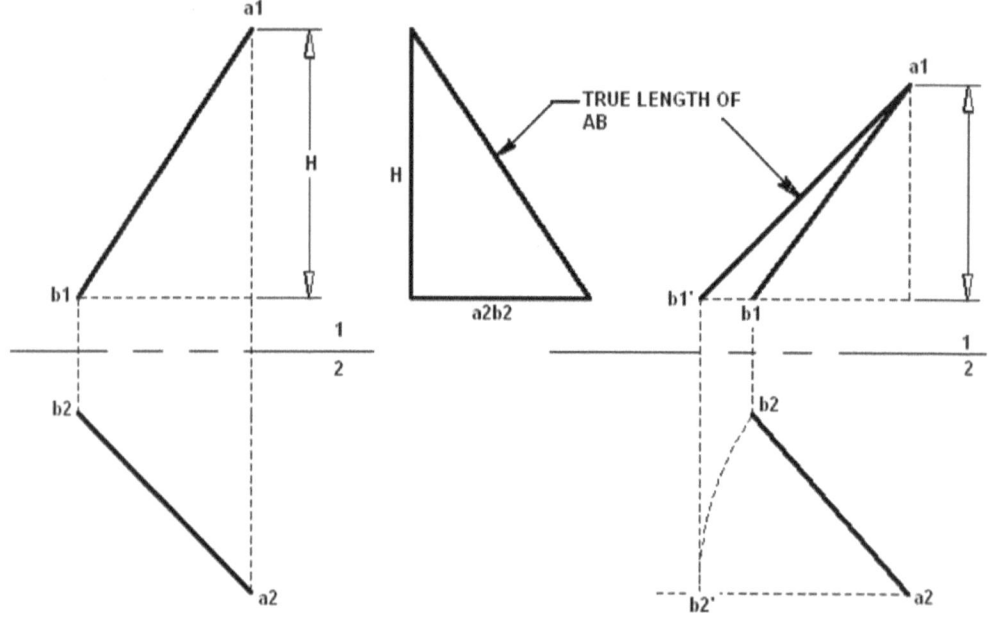

Fig. 7.7 True length of an oblique line by triangulation method

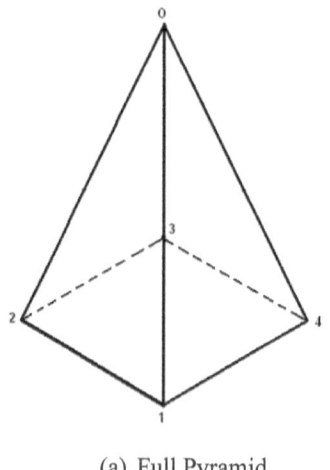

(a) Full Pyramid

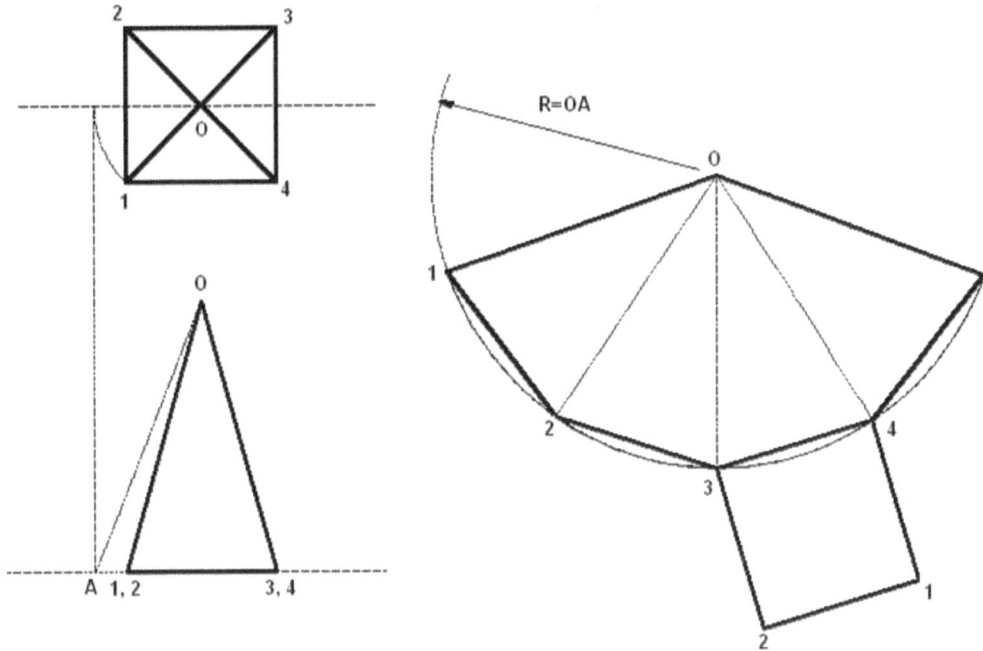

Fig. 7.8 Development of full pyramid

2. Development of a Truncated Pyramid

i.   Draw the front and top views of the truncated pyramid. Also, draw the auxiliary view of the inclined face.

ii.   Number the edges of the pyramid on the top view in the clockwise direction. Also, number the edges on the front view in agreement with the numbering on the top view.

iii.   Find the true lengths of all edges of the pyramid by triangulation method. Note that once the true length view OA of the edges of the full pyramid has been found, the other points B, C, etc. can be found horizontally projecting the end points of the edges from the front view.

iv.   With O as center and OA as radius draw arc.

v.   Use the true lengths of the sides of the base as radii to draw arcs that will intersect the arc drawn in step 4 at points 1, 2, 3, etc. as shown in figure below.

vi.   Join points 1, 2, 3, etc. to each other and to the center O of the arc drawn instep 4.

vii.   Transfer the true lengths of the edges from the true length diagram to the development, locating 1', 2', 3', etc.

177

viii.   Join 1', 2', 3', etc. with straight lines, and attach the base face and the inclined face to complete the development.

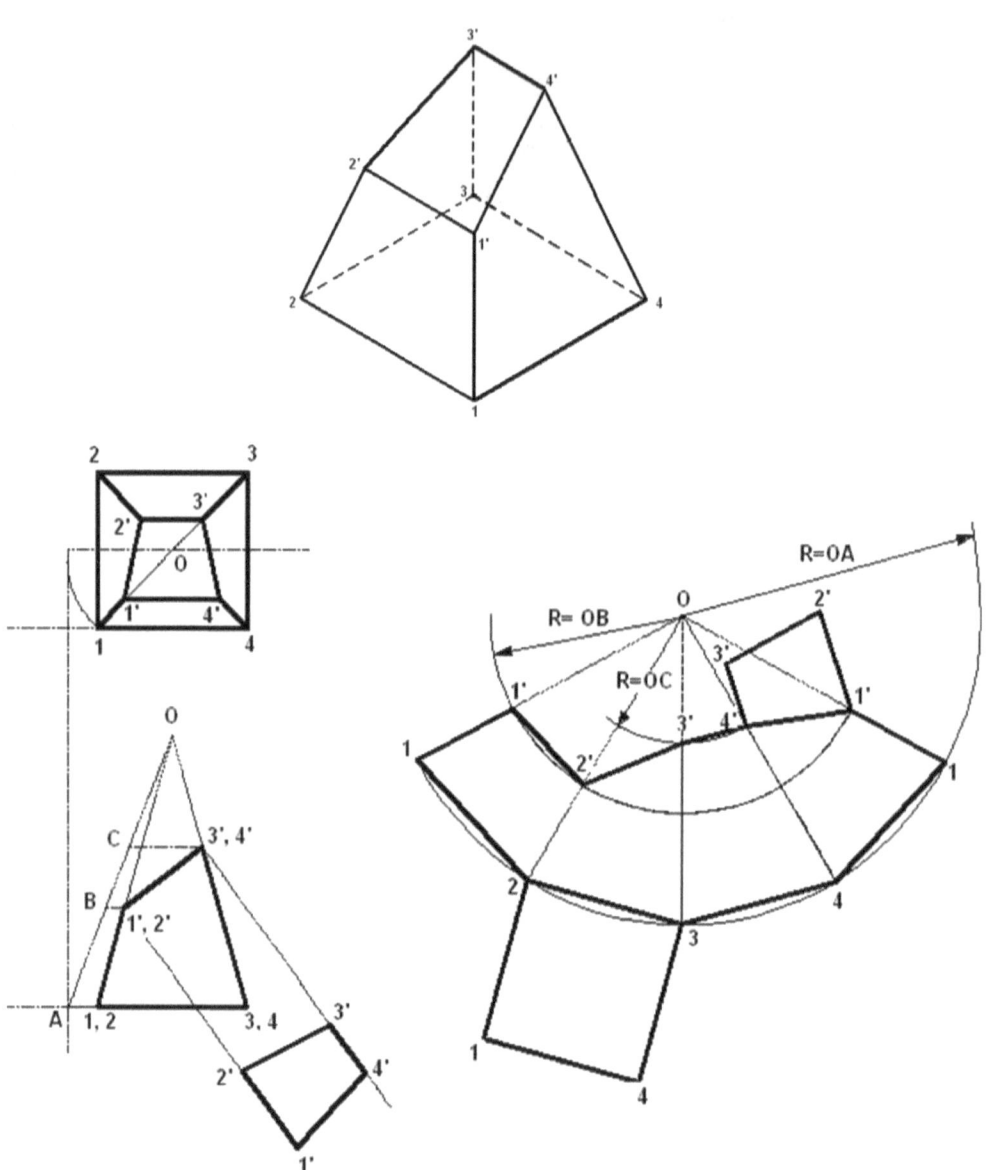

Fig. 7.9 Development of a Truncated Pyramid

178

(b) Development of Cones

The development of cones is also based on the radial development method. The cone may be considered as a pyramid that has infinite number of edges. Thus, the development of a cone may be made in a similar manner to that of the pyramid. The development of a right full cone is simply a sector whose radius is equal to the hypotenuse of the cone and whose arc length is equal to the circumference of the base of the cone. The subtended angle θ can be found using the formula:

$$\theta = \frac{r}{s} x360^o$$

Where r is the radius of the base of the cone and s is the hypotenuse of the cone.

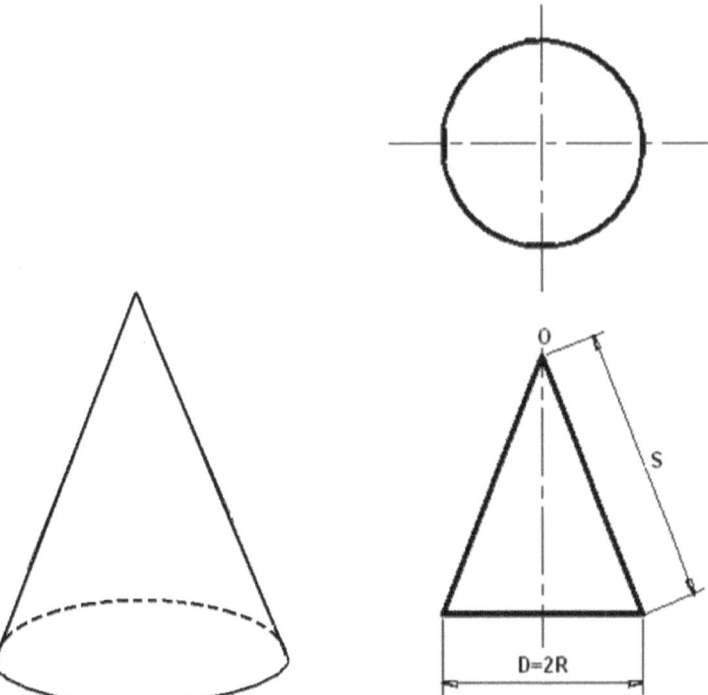

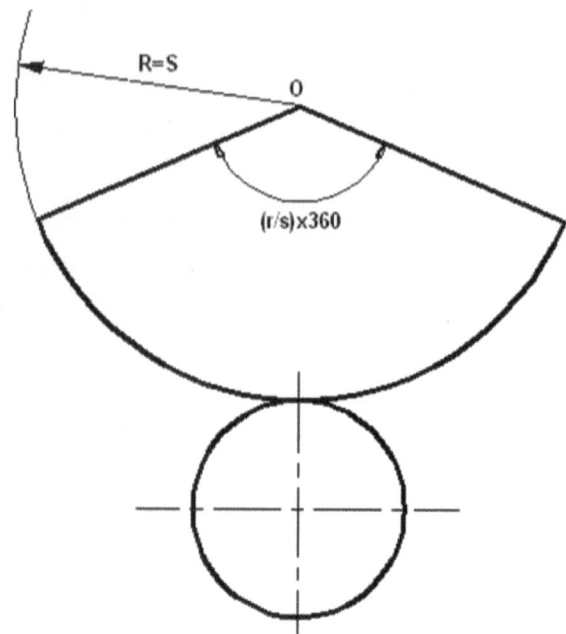

Fig. 7.10 Development of right cone

(i) Development of Truncated Cone

The intersection of a cone and a plane (that is neither perpendicular nor parallel to the axis of the cone and that is not parallel to the surface elements) is an ellipse. The true size of the ellipse is found by drawing its auxiliary view.

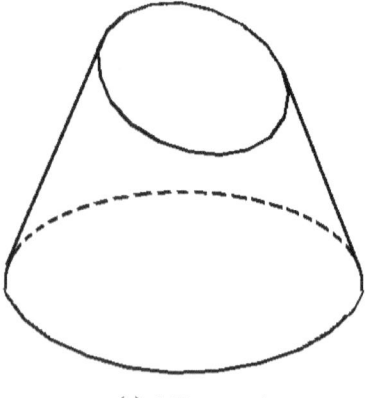

(a) A Truncated cone

Note that the top view of the ellipse is found as follows

1. Construct the front view of the truncated cone.
2. Extend the non-parallel lines to locate the apex of the full cone, and draw the top view of the full cone.
3. Divide the top view into any appropriate number of equally spaced surface line elements.
4. Project these surface elements to the front view, locating points a, b, c, etc. on the line representing the edge view of the ellipse.
5. Project points a, b, c, etc. to the top view on the corresponding line elements, locating a', b', c', etc.
6. Draw a smooth curve through points a', b', c', etc. to get the top view of the ellipse.

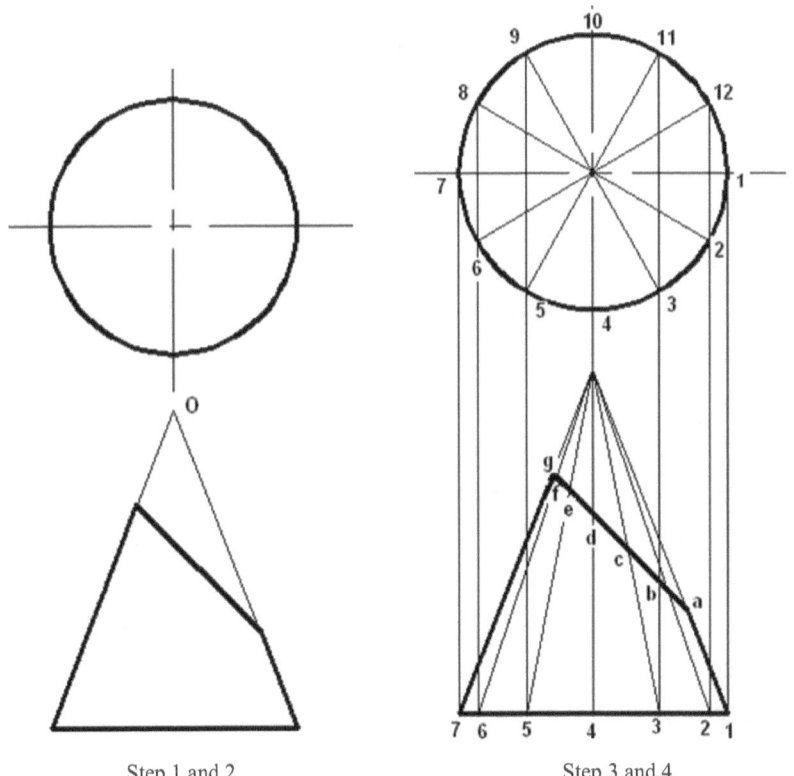

Step 1 and 2   Step 3 and 4

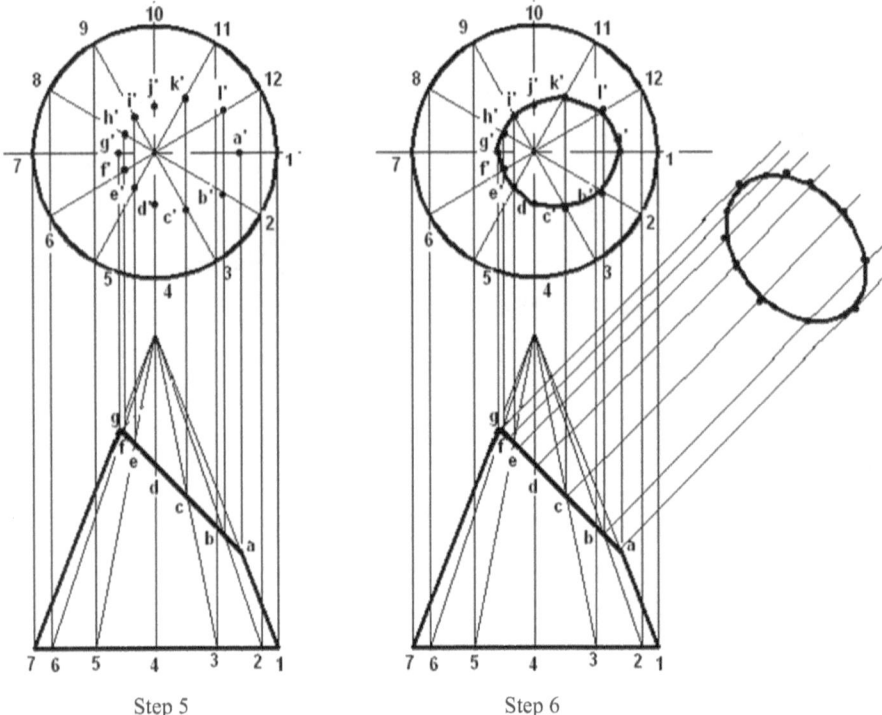

Step 5   Step 6

Steps for Development of Truncated cone

1. Draw the front and top views of the truncated cone. Also, draw the auxiliary view of the inclined surface, which is ellipse.

2. Divide the top view into any appropriate number of equal parts, say 12. Then, number them in the clockwise direction, as shown in figure below.

3. Project the surface line elements from the top view to the front view, and number them.

4. Project points from the edge view of the ellipse to the true length line O-1 to get the true lengths of all line elements.

5. Draw the development of the full cone. Then, divide into the same number of equal parts as the number of surface line elements taken in step 2 and mark points 1, 2, 3, etc.

6. Transfer the true lengths of all elements from the front view to the development, locating points a, b, c, etc.

7. Draw a smooth curve through points a', b', c', etc. and attach the base and the inclined surface (the ellipse) to complete the development.

182

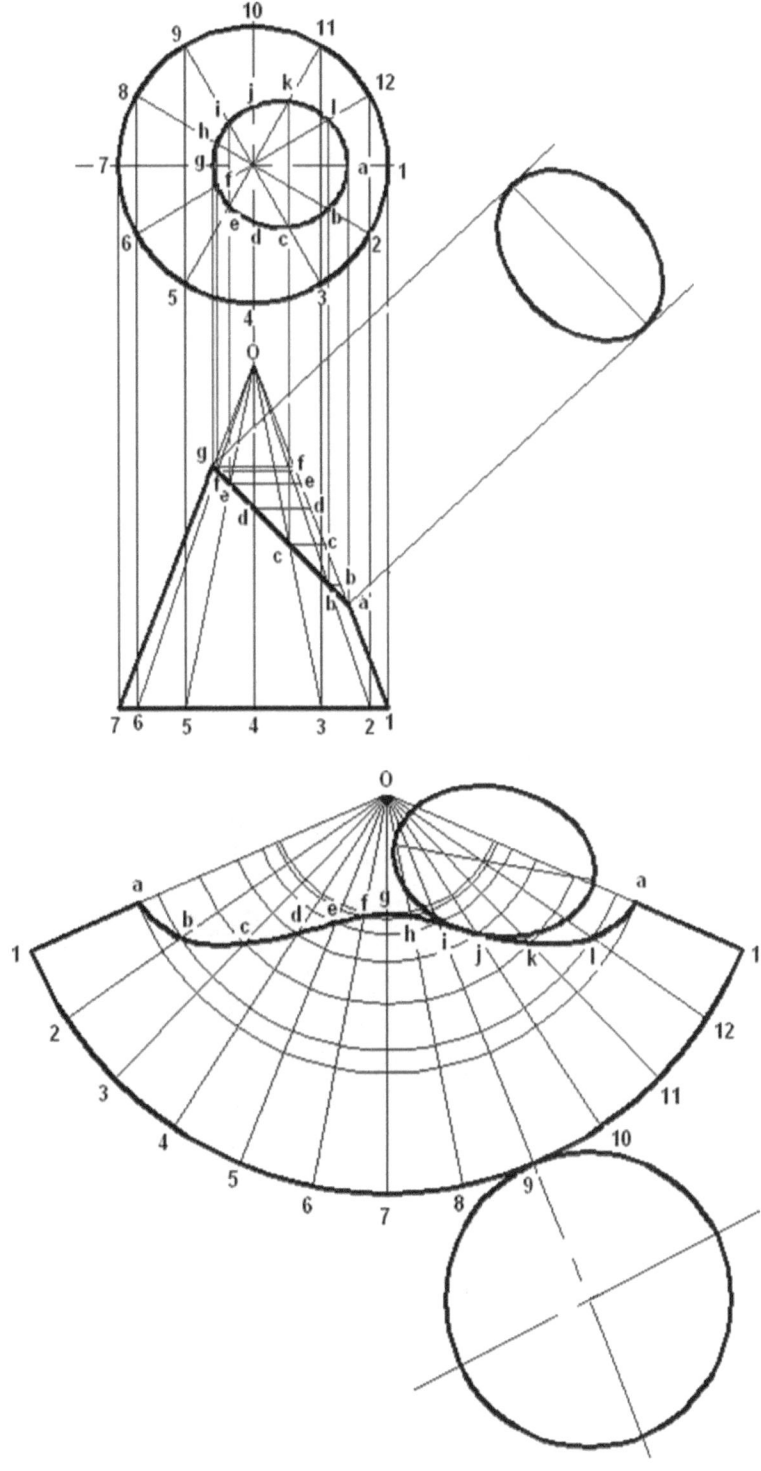

Fig. 7.11 Development of Truncated cone

7.2 Intersection of Surfaces

When two solids of the same or different cross-section come in contact each other, the surfaces of both the solids intersect one another. The two surfaces come in contact with one another and this sort of surface contact is known as "Intersection of surfaces of Solids".

The principle involved in intersections of planes and solids have their practical application in the cutting of openings in roof surfaces for flues and stacks, in wall surfaces for pipes, chutes, etc., and in the building of sheet-metal structures like tanks, boilers, etc. In such cases the problem is generally determining the true size and shape of the intersection of a plane and the geometric solid. Two surfaces that meet form a line of intersection. A pipe going through a wall is an example of an intersection. As an engineer, you must be able to determine the exact point of intersection between the pipe and the wall.

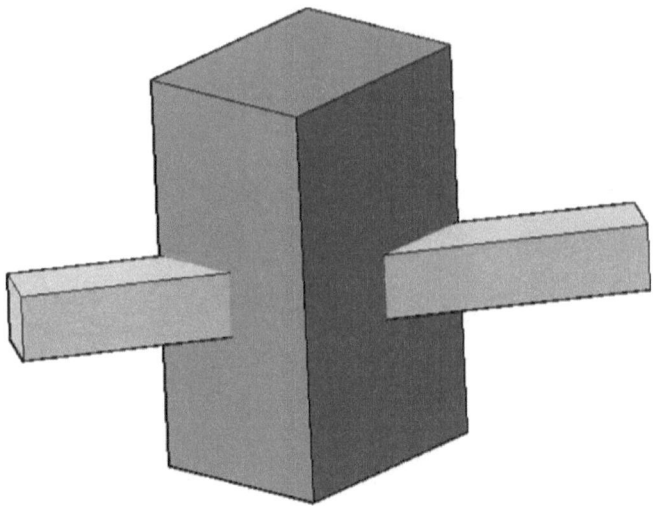

Fig. 7.12 Intersection of Prisms

When two solids intersect or interpenetrate each other, the line of intersection must be determined. Solids such as cylinders, prisms, pyramids, cones and spheres are the common intersecting bodies in sheet metal works.

Whenever two surfaces meet, there is a line common to both called the **line of intersection**. When making orthographic drawing of objects that comprise two or more intersecting parts, the line of intersection of these parts must be plotted on the orthographic views.

184

The line of intersection of two geometric surfaces is found by *determining a number of points common to both the surfaces and joining these points in correct order*. The resulting line may be straight, curved, or straight and curved depending upon the nature of the intersecting surfaces.

The intersecting surfaces may be broadly classified under three categories:

1. The intersection of plane surfaces,
2. The intersection of two curved surfaces, and
3. The intersection of a plane and curved surface.

When two surfaces bounded by planes, e.g., prism and prism, intersect each other the line of intersection is made up of straight lines. When two solids having curved surfaces, e.g., cylinder and cylinder, intersect each other the line of intersection is curve. Similarly, when two solids, one bounded by planes and the other a solid having a curved surface, e.g., prism and cylinder, intersect each other the line of intersection is a curve.

The geometric principles involved in intersection of surfaces have many practical applications in engineering, such as sheet metal work used for fabricating tanks, boilers, pipe joints, stacks, air conditioning ducts, chutes, bins, hoppers, etc.

7.2.1   Piercing point, Visibility and hidden line of intersection

(a) Visibility

Visibility, as used here, is the clear and correct representation of the relative positions of two geometric figures in multiview drawings. In each view, the visibility of the figures is indicated by drawing the figure that is in front (i.e., the most visible) with object lines entirely, while drawing the second figure with both object lines and dashed lines.

The dashed lines would be used for those features of the second figure that are behind the first figure and are therefore not visible (hidden) in that view.

The basic concept of visibility can be demonstrated with two skew lines, in which one line is either in front of, behind, above, below, to the left, or to the right of the other line.

Consider two skew lines AB and CD shown below

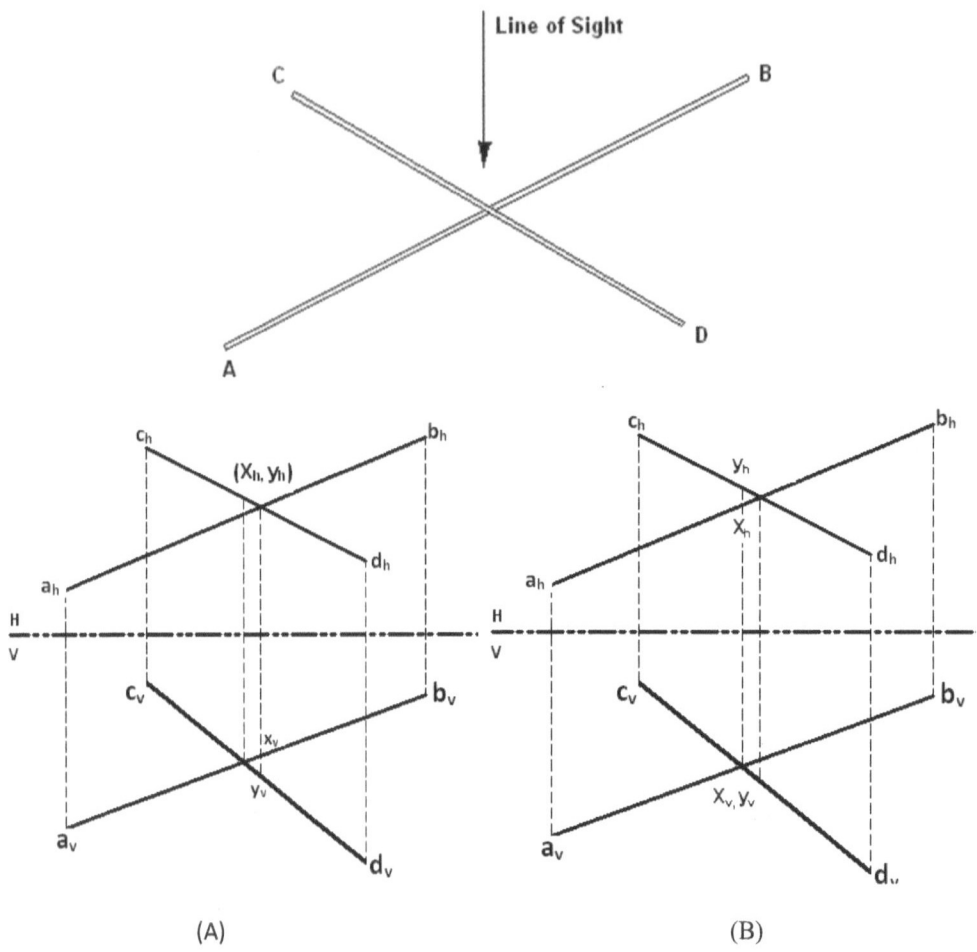

(A)   (B)

Rule of Visibility

To determine the visibility, let us assume that at the point of intersection in the top view x_h is on line AB and point y_h on line CD. From the front view, it can be seen that x_v is higher than y_v (nearer to the observer). Therefore, line $a_h b_h$ is behind $c_h d_h$.

In the front view of figure B above assume that point X_v is on line AB and point y_v is on line CD. But when the points are projected to the top view, it can be seen that X_h is below point Y_h. Therefore, line CD is above line AB.

(b) Piercing point

When a straight line intersects a plane, it pierces the plane at one point. The point is known as the **piercing point.**

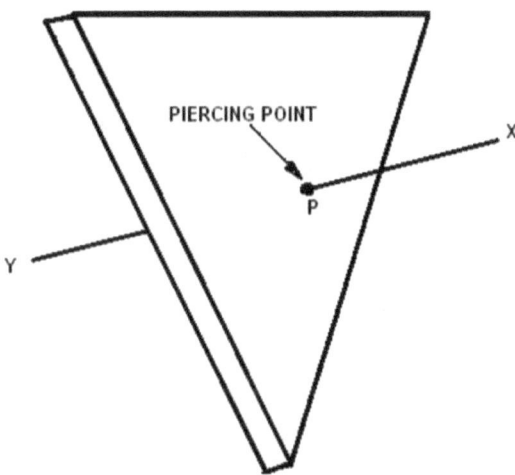

On orthographic views, the piercing point of a line and a plane can be found by one of the following methods:

    a) Cutting plane method

    b) Edge view method

(a) Cutting plane method

Consider line XY piercing plane ABC, if the vertical cutting plane MNX containing line XY is passed it will intersect the planes ABC on line WZ. The piercing point will be on the line XY and on the line of intersection WZ.

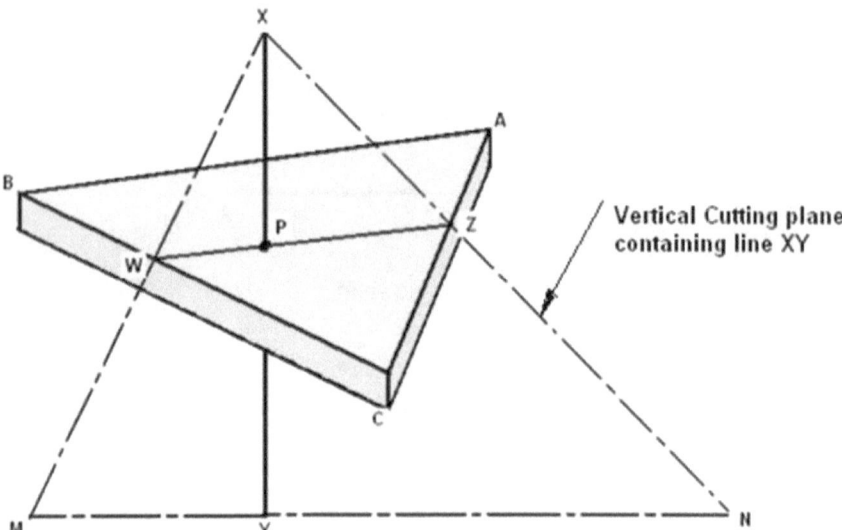

Vertical Cutting plane
containing line XY

The horizontal and vertical projection of planes XYZ and line LM are given on figure below. To find the piercing point of line LM with plane XYZ by cutting plane method follow the following steps.

1. Consider that $l_h m_h$ is the **edge view of the vertical cutting plane**, which will cut the plane at c_h and b_h respectively.

2. Draw projectors from c_h and b_h to obtain c_v and b_v on the respective edges of the plane $X_v Y_v Z_v$.

3. Join $c_v b_v$ and $l_v m_v$. The intersection point p_v is the piercing between the line and the plane in the front view.

4. Draw projector from p_v to obtain p_h on line $l_h m_h$, which is the piercing point for the top view.

5. Find the visibility between the line and the plane using the rule of visibility. Part of the line which is behind the plane is invisible and it is represented by hidden lines.

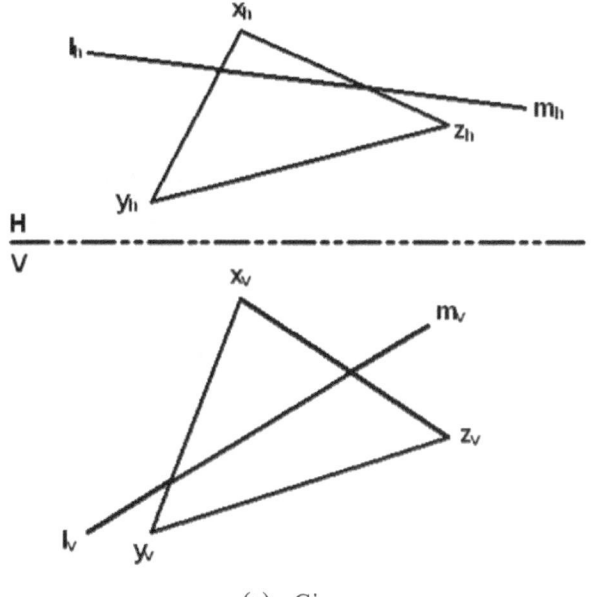

(a) Given

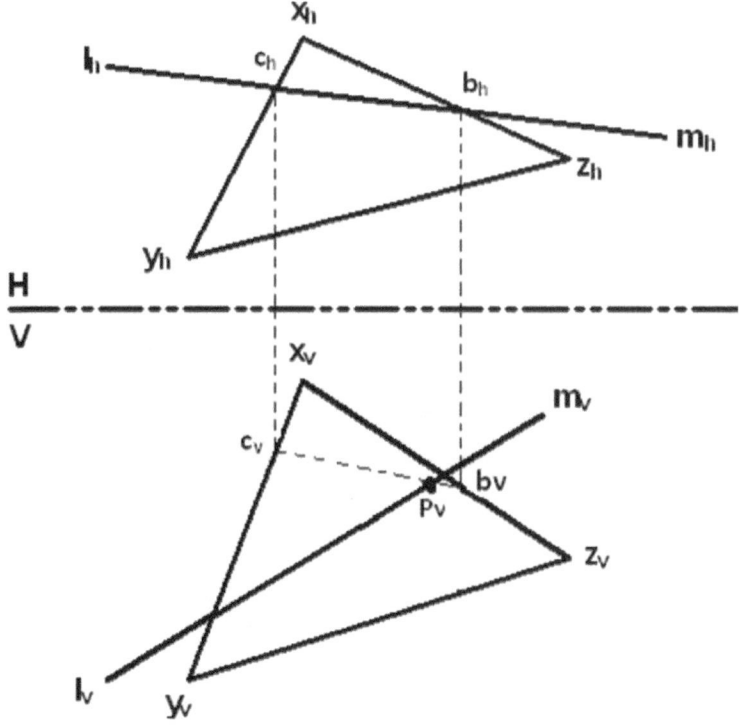

(b) Steps 1 to 3

189

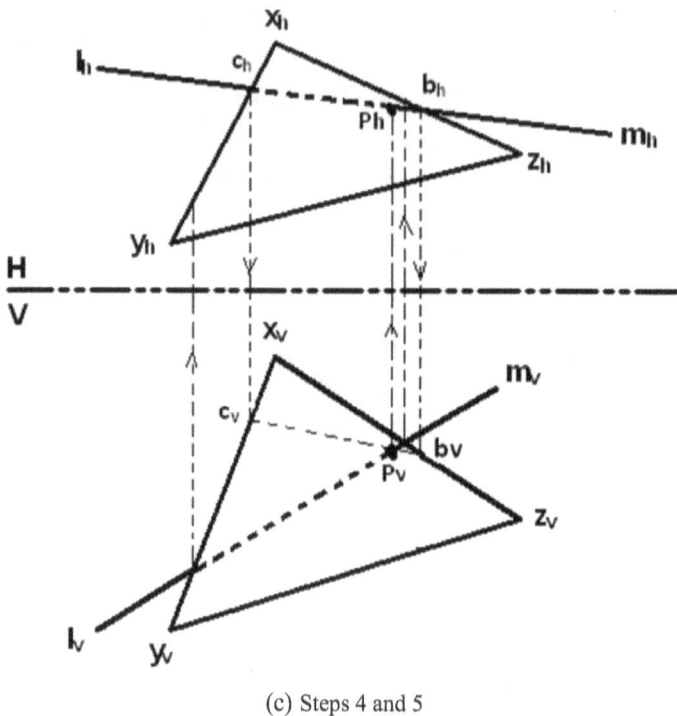

(c) Steps 4 and 5

7.2.2   Determining the intersection between a line and a plane: Edge view Method

The front and top views of plane ABC and Line LG are given: to find the piercing point of line LG with plane ABC by edge view method follow the following steps:

Step 1: Draw horizontal line from one of the vertices of the plane, A_FX_F.

Step 2: Find the true length of line AX on the adjacent projection plane, A_HX_H.

Step 3: Introduce an auxiliary projection plane 1 to find the edge view of plane ABC, and transfer distances from the front view to auxiliary plane 1.

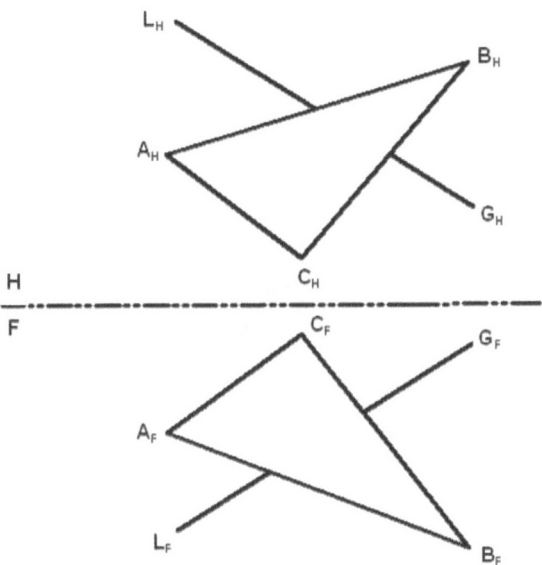

7.2.3 Intersection of Two Planes

The intersection of two planes is a straight line all of whose points are common to both planes. The line of intersection between two planes is determined by locating the piercing points of lines from one plane with the other plane and drawing a line between the points. The piercing points are located using either the edge view or cutting plane method.

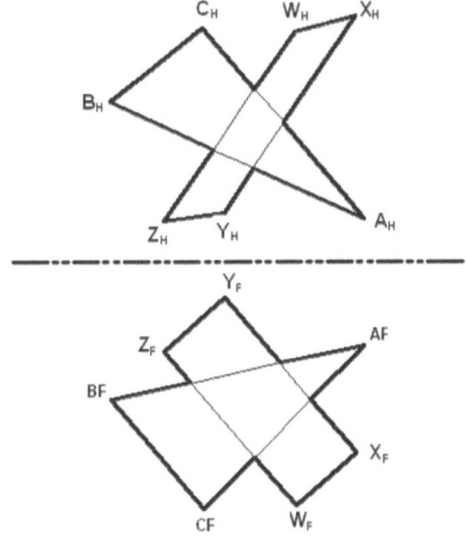

7.3 Problems

1. Make the development of a truncated cylinder shown below in Fig. 1
2. Make the development of the truncated hexagonal pyramid shown in Fig. 2
3. Make the development of the truncated cone shown in the Fig. 3

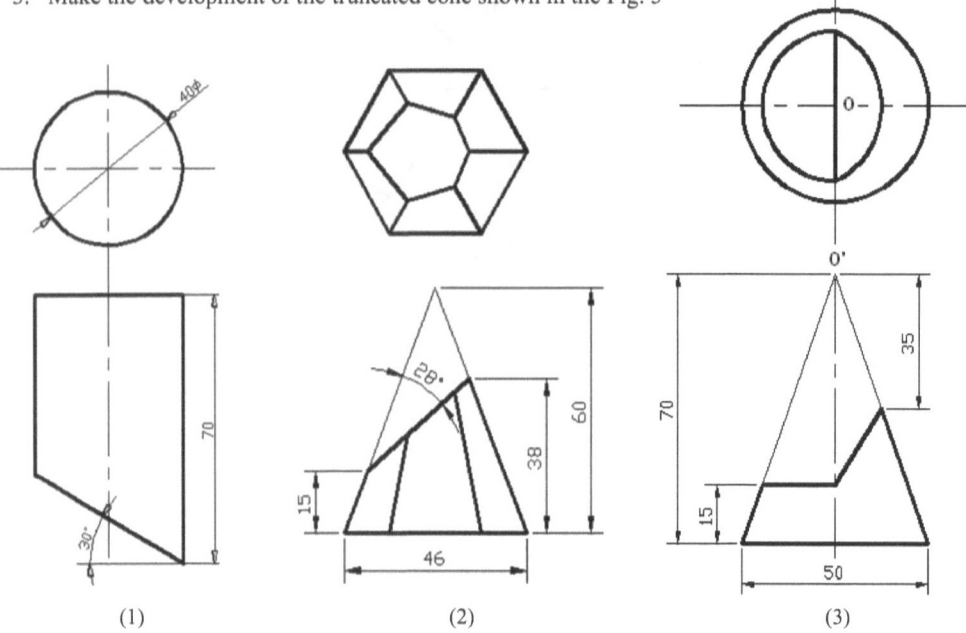

(1)   (2)   (3)

4. Draw the views of the truncated cone in fig. 4 below. Also, draw the auxiliary view of the inclined surface and make the development of the truncated cone.

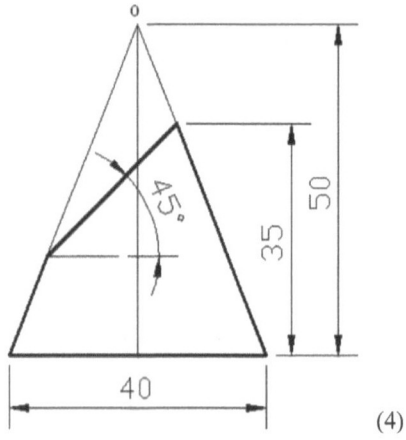

(4)

5. Front and top views of plane ABC and line LM are given. Find the piercing point of line and plane by cutting plane method and show the visibility on both views

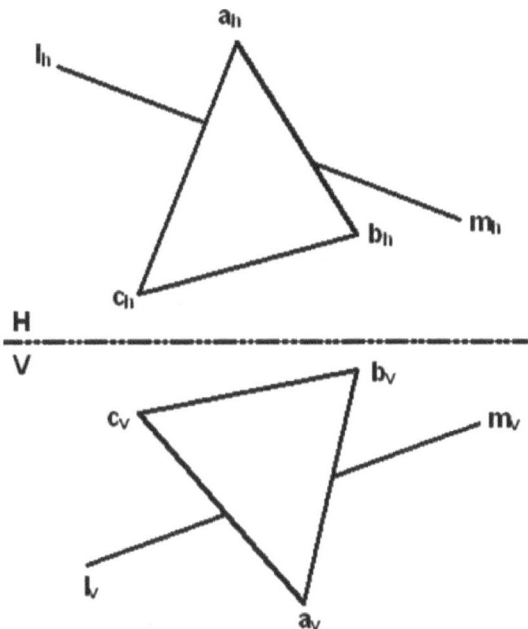

References

1. Bertoline, G. R., Wiebe, E. N., Hartman, N. W. & Ross, W. A., 2009. *Technical Graphics Communications*. Fourth Edition ed. New York: McGraw-Hill.

2. Bertoline-Wiebe, 2007. *Engineering Graphics. Fundamentals of Graphics Communication*. Fifth Edition ed. New York: McGraw-Hill.

3. Jensen, C. & Helsel, J. D., n.d. *Engineering Drawing and Design*. s.l.:s.n.

4. Madsen, D. A. & Madsen, D. P., 2012. *Engineering Drawing & Design*. Fifth Edition ed. New York: Cengage Learning.

5. R.S, V., n.d. *Engineering Drawing and Graphics*. s.l.:s.n.

6. Reddy, K. V., 2008. *Text Book of Engineering Drawing*. Second Edition ed. Hyderabad: BS Publications.

194

www.ingramcontent.com/pod-product-compliance
Lightning Source LLC
Chambersburg PA
CBHW030625220526
45463CB00004B/1421